Chemistry Research and Applications

Chemistry Research and Applications

What to Know about Lanthanum
Catherine C. Bradley (Editor)
2023. ISBN: 979-8-88697-615-1 (Softcover)
2023. ISBN: 979-8-88697-623-6 (eBook)

The Future of Biorefineries
Waldemar Nyström (Editor)
2023. ISBN: 979-8-88697-524-6 (Hardcover)
2023. ISBN: 979-8-88697-528-4 (eBook)

Properties and Uses of Antimony
David J. Jenkins (Editor)
2022. ISBN: 979-8-88697-081-4 (Softcover)
2022. ISBN: 979-8-88697-088-3 (eBook)

The Science of Carbamates
Güllü Kaymak (Editor)
2022. ISBN: 978-1-68507-708-2 (Softcover)
2022. ISBN: 978-1-68507-872-0 (eBook)

Deep Eutectic Solvents: Properties, Applications and Toxicity
Carlos Eduardo de Araújo Padilha, PhD, Everaldo Silvino dos Santos, PhD, Francisco Canindé de Sousa Júnior, PhD, Nathália Saraiva Rios, PhD (Editors)
2022. ISBN: 978-1-68507-719-8 (Hardcover)
2022. ISBN: 978-1-68507-799-0 (eBook)

Polycyclic Aromatic Hydrocarbons: Sources, Exposure and Health Effects
Warren L. Gregoire (Editor)
2022. ISBN: 978-1-68507-626-9 (Softcover)
2022. ISBN: 978-1-68507-685-6 (eBook)

More information about this series can be found at
https://novapublishers.com/product-category/series/chemistry-research-and-applications/

Charles R. Howe
Editor

Pyrene

Chemistry, Properties and Uses

Copyright © 2023 by Nova Science Publishers, Inc.

All rights reserved. No part of this book may be reproduced, stored in a retrieval system or transmitted in any form or by any means: electronic, electrostatic, magnetic, tape, mechanical photocopying, recording or otherwise without the written permission of the Publisher.

We have partnered with Copyright Clearance Center to make it easy for you to obtain permissions to reuse content from this publication. Simply navigate to this publication's page on Nova's website and locate the "Get Permission" button below the title description. This button is linked directly to the title's permission page on copyright.com. Alternatively, you can visit copyright.com and search by title, ISBN, or ISSN.

For further questions about using the service on copyright.com, please contact:
Copyright Clearance Center
Phone: +1-(978) 750-8400   Fax: +1-(978) 750-4470   E-mail: info@copyright.com

NOTICE TO THE READER

The Publisher has taken reasonable care in the preparation of this book, but makes no expressed or implied warranty of any kind and assumes no responsibility for any errors or omissions. No liability is assumed for incidental or consequential damages in connection with or arising out of information contained in this book. The Publisher shall not be liable for any special, consequential, or exemplary damages resulting, in whole or in part, from the readers' use of, or reliance upon, this material. Any parts of this book based on government reports are so indicated and copyright is claimed for those parts to the extent applicable to compilations of such works.

Independent verification should be sought for any data, advice or recommendations contained in this book. In addition, no responsibility is assumed by the Publisher for any injury and/or damage to persons or property arising from any methods, products, instructions, ideas or otherwise contained in this publication.

This publication is designed to provide accurate and authoritative information with regard to the subject matter covered herein. It is sold with the clear understanding that the Publisher is not engaged in rendering legal or any other professional services. If legal or any other expert assistance is required, the services of a competent person should be sought. FROM A DECLARATION OF PARTICIPANTS JOINTLY ADOPTED BY A COMMITTEE OF THE AMERICAN BAR ASSOCIATION AND A COMMITTEE OF PUBLISHERS.

Additional color graphics may be available in the e-book version of this book.

Library of Congress Cataloging-in-Publication Data

ISBN: 979-8-88697-670-0

Published by Nova Science Publishers, Inc. † New York

Contents

Preface vii

Chapter 1   **Analytical Applications of Pyrene in Forensic Science**..........................1
Kirti Kumari Sharma, Tulsidas R. Baggi and V. Jayathirtha Rao

Chapter 2   **Pyrene as a Chromophore in Various Organic Materials**........................35
V. Jayathirtha Rao

Chapter 3   **The Presence of Micro and Nanoplastics Influencing Pyrene Toxicity**...........77
Hillary Henao-Toro, Edwin Chica and Ainhoa Rubio-Clemente

Chapter 4   **Self-Assembling Supramolecular Structures of Pyrene**.........................97
Sandeep Kumar

Bibliography 117

Index 139

Preface

This book contains five selected chapters on the chemistry, properties, and uses of pyrene. Chapter One reviews the analytical applications of pyrene in forensic science. Chapter Two examines pyrene as a chromophore in various organic materials applications. Chapter Three reviews how the presence of micro and nanoplastics can influence pyrene toxicity. Chapter Four looks at the self-assembling supramolecular structure of pyrene.

Chapter 1 - Pyrene is the smallest peri-fused polycyclic aromatic hydrocarbon. The pyrene molecule is highly symmetrical (point group D2h) and despite having 16 π electrons and not following the Huckle's 4n + 2 rule, it is aromatic. These particular features provide to pyrene interesting electronic properties, which are highly exploited in the characterization of host–guest systems. Without interactions with potential host molecules and in diluted solutions to avoid excimeric formations, pyrene presents in solution an intense and anisotropic fluorescence, as well as a high fluorescence quantum yield. Pyrene is a strong electron donor material and can be combined with several materials in order to make electron donor-acceptor systems which can be used in energy conversion and light harvesting applications. Pyrene chromophore, because of its UV-Visible absorption, fluorescence and excimer luminescence properties is being extensively used in forensic applications. This chapter is aimed at bringing the information on the utility of pyrene as chromophore in various forensic applications like: (i) Sensing Explosives; (ii) Finger marks (iii) Bio imaging and (iv) Ion Sensing. Publications related to above topics will be briefed to highlight the contribution of pyrene and its optical properties.

Chapter 2 - Pyrene was identified as one of the aromatic substances from the coal-tar distillation. This aromatic residue, not obeying [4n+2] Huckel's rule, has been subjected to a variety of research investigations because of its interesting properties. Notably, its light absorption, fluorescence light emission, excimer luminescence, medium polarity sensitive luminescence, quenchable fluorescence, aromatic stacking nature and DNA interactable fluorophore properties are extensively utilized for making innumerable substrates having a variety of optical and optoelectronic applications. This chapter is focused on narrating some basic (i) photophysical properties of pyrene and its derivatives; (ii) pyrene as a core molecule in OLED materials; (iii) pyrene as a molecule in organic solar cell materials; and (iv) pyrene in Non-Linear Optical properties.

Chapter 3 - Pyrene is a polycyclic aromatic hydrocarbon with carcinogenic, mutagenic and teratogenic potential. In turn, micro and nanoplastics have been recently known as contaminants of emerging concern, whose detection and quantification in hydric resources is increasing every day. Due to the physicochemical properties of pyrene, especially the low solubility in water and the high affinity for particulate matter, this substance has a tendency to be sorbed into solid materials such as micro and nanoplastics, which are extensively distributed worldwide. In this work, the influence of the presence of micro and nanoplastics in particularly aquatic ecosystems on pyrene toxicity is analyzed. It was evidenced that even though micro and nanoplastics delayed the appearance of pyrene potential detrimental effects on biota, a synergic effect can occur, leading to drastic consequences on aquatic fauna and flora biodiversity. In this regard, the co-existence of pyrene and micro and nanoplastic particles in the environment need to be urgently addressed so that the biota risks associated are minimized.

Chapter 4 - Self-assembling supramolecular structures formed by functionalized disc-shaped molecules are commonly known as discotic liquid crystals (DLCs). Upon appropriate substitution, pyrene derivatives exhibit liquid crystalline properties. Pyrene as a core for DLCs was discovered in 1995. During the past three decades, a number for liquid crystalline derivatives of pyrene have been synthesized and characterized. In this chapter, the synthesis, characterization and physical properties of some self-assembling supramolecular structures of pyrene core are presented.

Chapter 1

Analytical Applications of Pyrene in Forensic Science

Kirti Kumari Sharma[1], Tulsidas R. Baggi[2*] and V. Jayathirtha Rao[1,*]

[1]Fluoro Agro Chemicals Department and AcSIR-Ghaziabad,
CSIR-Indian Institute of Chemical Technology, Uppal Road Tarnaka,
Hyderabad, TS, India
[2]Formerly of Central Forensic Science Laboratory, Hyderabad,TS, India

Abstract

Pyrene is the smallest peri-fused polycyclic aromatic hydrocarbon. The pyrene molecule is highly symmetrical (point group D2h) and despite having 16 π electrons and not following the Huckle's 4n + 2 rule, it is aromatic. These particular features provide to pyrene interesting electronic properties, which are highly exploited in the characterization of host–guest systems. Without interactions with potential host molecules and in diluted solutions to avoid excimeric formations, pyrene presents in solution an intense and anisotropic fluorescence, as well as a high fluorescence quantum yield. Pyrene is a strong electron donor material and can be combined with several materials in order to make electron donor-acceptor systems which can be used in energy conversion and light harvesting applications. Pyrene chromophore, because of its UV-Visible absorption, fluorescence and excimer luminescence properties is being extensively used in forensic applications. This chapter is aimed at bringing the information on the utility of pyrene as

---

* Corresponding Author's Email: trbaggi@gmail.com.

In: Pyrene: Chemistry, Properties and Uses
Editor: Charles R. Howe
ISBN: 979-8-88697-670-0
© 2023 Nova Science Publishers, Inc.

chromophore in various forensic applications like: (i) Sensing Explosives; (ii) Finger marks (iii) Bio imaging and (iv) Ion Sensing. Publications related to above topics will be briefed to highlight the contribution of pyrene and its optical properties.

Introduction

In literature, many fluorophores are reported such as anthracene, naphthalene [1], coumarin- 6 [2], rhodamine [3–5], fluorine linked phenanthro imidazole [6], fluorescent nanoparticles [7–11]. Pyrene possesses comparatively higher quantum yield than anthracene, naphthalene and rhodamine B. Pyrene is an especially useful fluorophore due to its well-resolved absorption and emission spectra, long lifetime in the excited singlet state, ability to form excimers with a distinct fluorescence, and ability to measure the polarity of its molecular environment [12]. Pyrene and its derivatives find their application in commercial dyes and dye precursors. Pyrene based D–π–A dyes exhibit solvatochromism and high fluorescence in non-polar solvents and water [13] having good absorption and emission properties.

Pyrene is the smallest peri-fused polycyclic aromatic hydrocarbon and is obtained during the combustion of organic compounds. The pyrene molecule is highly symmetrical (point group D2h) and despite having 16 π electrons and not following the Hückel's 4n + 2 rule, it is aromatic. These particular features provide to pyrene interesting electronic properties, which are highly exploited in the characterization of host–guest systems. Without interactions with potential host molecules and in diluted solutions to avoid excimeric formations, pyrene presents in solution an intense and anisotropic fluorescence, as well as a high fluorescence quantum yield. The sensitivity of pyrene, to the polarity of the environment is a photo physical property that is extensively used to study host–guest interactions. Pyrene was the first molecule for which excimer behavior appearing at 450 nm was discovered. This photochemical property of pyrene is its excimer emission in more concentrated solutions. The ground-state dimer formation of pyrene molecules leads to a large characteristic band at ca. 480 nm, which is of interest in determining a stacking of two or more pyrene molecules in solution. Pyrene is a strong electron donor material and can be combined with several materials in order to make electron donor-acceptor systems which can be used in energy conversion and light harvesting applications. Polycyclic aromatic compounds based on pyrene are often useful as a chromophore and can also serve as semiconductor in (opto-) electronic devices. The extension of the so-called K-Region of pyrene provides various π-conjugated systems with useful properties.

Pyrene and its derivatives have been extensively used in several biochemical, chemical and forensic analytical applications such as detection / determination of nitro-aromatics, explosives, proteins, RNA/ DNA, PAHs, cyanide, aluminum and aptamer based detection of complex molecules in biological matrices, petroleum products, lysozyme, fingermarks (fingerprints), potassium, and target ligands etc.

Pyrene is one of the best fluorophores showing optimized photo-physical characteristics such as high fluorescence quantum yield, long fluorescence lifetime and chemical stability making it ideal for ratiometric chemosensors. Further, pyrene exhibits monomer–excimer dual fluorescence, and the fluorescence intensity ratio of the excimer to monomer emission is sensitive enough for conformational changes of the pyrene sensor systems. Many pyrene-based ratio metric chemo sensors have been developed for many cations such as Hg^{2+}, Zn^{2+}, Pb^{2+}, Ag^+, Cu^{2+} and Al^{3+}. However, only a few sensors are developed for anions such as F^- and HSO_3^-.

The AIE-active pyrene conjugates appear to be promising candidates in bio imaging studies. Pyrene is an aromatic hydrocarbon with a polycyclic flat structure containing four fused benzene rings to provide an unusual electron delocalization feature that is important in the AIE property. Numerous pyrene-based AIE-active materials have been reported with the AIE property towards sensing and imaging applications. Most importantly, these AIE-active pyrene moieties exist as small molecules, Schiff bases, polymers, supramolecules, metal-organic frameworks, etc.

Explosives / Nitro Aromatics

Since the nitro-compounds (NCs) are the main constituents of explosives, the detection of these materials is most desirable in a variety of forensic and environmental applications such as in health care, mine fields, blast sites, waste water treatment, industrial processes etc., where timely detection of explosives could ensure safety and security of mankind. Such applications have gained considerable significance in wake of the imminent terror threats faced by the world. Among the current methods for the detection of explosives, fluorescence based chemical sensors have attracted a considerable interest owing to their simplicity, high sensitivity, easy visualization, and real time monitoring. Among the fluorescent probes, pyrene is widely used as an efficient analytical tool in detecting toxic nitric oxide (NO) in NO-releasing oxygen-sensing polymer films, measuring intracellular molecular concentrations, as well as in the detection of electron deficient species and the explosives. Most of the pyrene based sensors exhibit well defined monomer emission at 370–420 nm and an excimer emission at 480 nm of the pyrene units, subject to their molar concentration in solution. A variety of electron deficient species are able to quench their fluorescence emission via charge-transfer or electron-transfer mechanisms leading to excellent detection of electron deficient molecules, including the samples containing explosives.

Singla P. et al. [14] have reported the detection and quantification of nitro aromatics of forensic importance based on the quenching of the excimer emission of pyrene moiety of the compound phenanthro[4,5 fgh]pyrido[2,3-b]quinoxaline Figure 1 structure (1). The compound 1 was synthesized by them carrying out Schiff's-base condensation between pyrene-4, 5-dione and 2, 3- diaminopyridine. A solution of 1 in THF showed excimer emission at 487 nm (K exc. 330 nm). They point out that the structure (1) can exist in closely spaced form in solution with shortest distance of 5-8 Å between the molecular planes of the two pyrene units enabling to form an excimer. The excimers were stable in the presence of Hg^{2+} ions and the emission of structure (1) gets quenched and was able to discriminate between various nitro aromatics based on their respective emission quenching efficiencies. The quenching observed of the excimer emission of structure (1) was able to discriminate between the derivatives of various nitro aromatics such as explosives which are electron deficient in nature. The authors suggest that suitable sensors could be developed for detecting and determining the explosives even at a trace level in various matrices. These workers tested the fluorescence behavior of structure (1) in the presence of several high energetic compounds such as 2,4-dinitroaniline (2,4- DNA), p-nitroaniline (p-NA), p-nitrophenol (p-NP), 2,4,6-trinitrophenol (picric acid, PA), 2,4-dinitrophenol (2,4-DNP), o-nitroaniline (o-NA), m-nitroaniline (m-NA), p-nitro benzoic acid (p-NBA), 1,4-dinitrobenzene (1,4-DNB), 2,4-dinitrotoluene (2,4-DNT), nitrobenzene (NB), o-chloro nitrobenzene (o-CNB), and m-chloro nitrobenzene (m-CNB), with nitro methane (NM), for comparison, which did not show any quenching effect. It was found that significant quenching of fluorescence emission of structure (1) was determined only in case of 2,4-DNA, p-NA, p-NP, PA, and 2,4-DNP, in the order 2,4-DNP < PA < p-NP < p-NA < 2,4-DNA, while NM did not show any significant fluorescence quenching. To check the applicability of structure (1) as a sensor, the authors used filter papers impregnated with solution of structure (1) was tested with trace levels of various NCs with naked eye and with UV fluorescence. The detection limits were reported as low as 9.1, 13.8, 13.9, 22.9, and 92 nanograms for 2, 4-DNA, p-NA, p-NP, PA, and 2, 4-DNP, respectively.

Conjugated oligopyrene based sensors with high fluorescence quantum yield, with "molecular wire" effect were developed for certain nitro aromatics by Hua Bai et al. [15]. They synthesized the conjugated oligopyrene by electro polymerization of pyrene in a mixed electrolyte of boron trifluoride diethyl etherate and diethyl ether. By spin coating they prepared thin film fluorescent sensors, which showed rapid fluorescence quenching when exposed to nitro aromatic compounds like tri nitro toluene, di nitro toluene and nitrobenzene. The efficiency of these sensors was attributed to large free energy change for electron transfer from the excited oligopyrene to the nitro aromatic molecules, and the twist chain structures in the sensing films. The authors give the detailed procedure for preparation of oligopyrene and pyrine films on quartz substrate by spin coating method. After the film is exposed to analyte vapors they were transferred to a fluorescence spectrometer to read at an excitation wavelength of 350 nm and 337 nm for oligopyrene and pyrene respectively. They mainly appear to have concentrated on the detection of TNT.

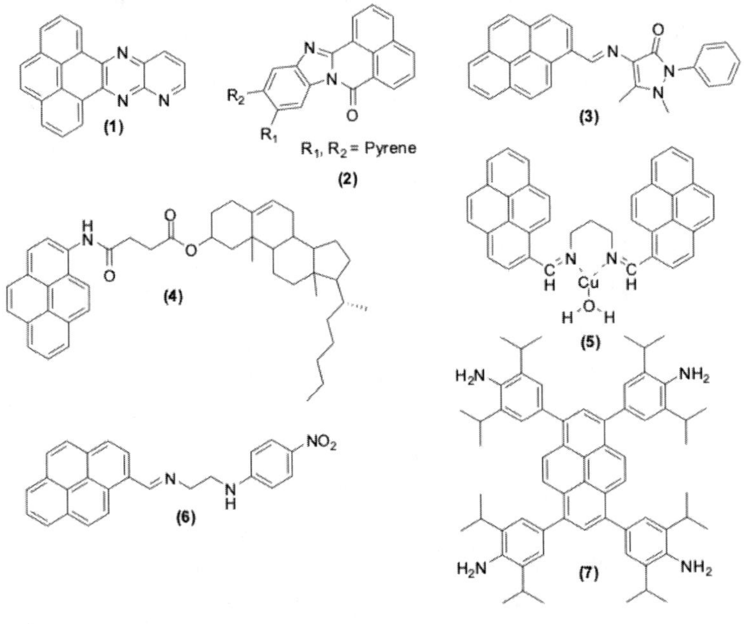

Figure 1. Differently substituted pyrene moieties for explosive/ nitro aromatics sensing.

Boonsri M. et al. [16] have developed a fluorescent on-off sensor for TNT based on aggregation induced emission enhancement of pyrenyl benzimidazol-isoquinolinones Figure 1 structure (2). They synthesized three new pyrene substituted benzimidazole-isoquinlinones through imidation condensation reaction and Suzuki coupling. These compounds show good aggregation- induced emission enhancement in aqueous THF media exhibiting selective fluorescent quenching towards TNT. They conclude saying that the compound with two pyrene units shows excellent selectivity and good quenching efficiencies. Compound containing two pyrene groups exhibits the best selective fluorescent quenching by TNT with the Stern-Volmer constant of 6 ´ 104 M^{-1} nd. The paper sensors prepared from the compound, with two pyrene units, can detect TNT visually both in aqueous and vapor phase, with naked eye with a detection limit of 0.25 ppm and at a lowest concentration of 50 μM.

Kathy- Sarah Focsaneanu and J. C. Scainano [17] demonstrate the utility of the sensors developed by them based on the differential monomer to excimer fluorescence ratio for determining electron deficient molecules such as explosives. The method is said to be suitable for rapid screening of explosives in different matrices. The procedure uses the quenching of the monomer and excimer emissions from pyrene for detection and determination of electron deficient molecules including explosives. The authors have used HPLC with a post column diode array detector, using the FM/FE ratios. They have used in this work steady state and time-resolved emission techniques to assess pyrene fluorescence quenching as a way to discriminate between electron-donor and electron-acceptor quenchers. They propose that the discrimination of these quenchers between monomer and excimer quenching can be used for explosive detection. This two-wavelength analysis of quenching of pyrene fluorescence demonstrates that the monomer-to-excimer fluorescence ratio senses explosives and other electron deficient molecules giving semi quantitative results (±15%) in the absence of authentic samples. The limit of detection is reported to be 10 ppm. The authors record that the method may sometimes give false positives with other electronegative compounds but in most of the cases it is unlikely to give false positives by detecting electron deficient molecules other than explosives.

Thakarda J. et al. [18] have synthesized a new luminescent gold nano structures in which 1-Pyreneiodide is attached on the surface of Au:PVP cluster. PY-Au:PVP gives strong characteristic emission being given from excimer emission. The method can determine nitro aromatic explosives by employing the 1- Pyreneiodide –ligated luminescent gold nano structures, through quenching of the monomer and excimer emissions. In this system the poorly luminescent 1-pyreneiodid exhibits intense emission when coupled with Au: PVP, despite the presence of a strong heavy atom effect. Detection of nitro aromatic explosives in solution and vapor phase at a nano molar level has been made possible through luminescence quenching with the new nano cluster developed by them. The authors also claim that the method can also be used for developing latent fingerprints. They have given the mechanism of luminescence quenching as static quenching arising from the supramolecular complexation between the electron-rich pyrenes and the electron-deficient nitro aromatic explosives. The work shows the application of the developed Au-nano clusters as novel "functional nano materials" with possible applications in forensic work for the detection of explosive traces and for developing latent fingerprints.

Taudte R. V. et al. [19] have developed a new innovative, simple, rapid and cost effective technique for the detection of explosives based on fluorescence quenching of pyrene on paper strip micro pad devices (mPADs) with wax barriers with optimized magenta color. One microliter of 0.5 mg/mL 21 pyrene dissolved in an 80: 20 methanol–water solution was deposited on the hydrophobic circle (5 mm diameter) to produce the active microchip device. Using UV radiation (365 nm) ten different organic explosives could be detected using these mPADs. The limits of detection ranged from 100–600 ppm. The authors have also developed a portable battery operated sensor based on the above principle for field use.

4-((pyren-1- yl) methylene amino)-1,2-dihydro-2,3-dimethyl-1-phenylpyrazol-5-one (PAP) Figure 1 structure (3) was synthesized by Milan Shyamal et al. [20] as an aggregate /solid state fluorescence compound for sensing traces of 2,4,6-trinitrophenol. PAP showed aggregation induced emission (AIE) providing cyan emissive fluogenic characteristic aggregates of 0-D to 1-D microcrystals which show optical waveguide effect. They studied the AIE mechanism in solid state by using dynamic light scattering, time resolved photoluminescence, optical fluorescence and scanning electron microscopy. They saw triple excitation induced triple emissions of green, yellow and red lights from the microstructures in solid state covering the entire visible spectral region. By using this AIE phenomenon and the hydrosol as a fluorescent chemo sensor they were able to sense the explosive 2, 4, 6 trinitro phenol (TNP), with a detection limit of 16 nM with excellent Stern-Volmer quenching constant of 4.7×10^5 M^{-1}. The authors successfully developed fluorescent paper strips to detect with naked eye, trace levels of TNP in solid state. The method is fast, cheap, reliable and proven for site visualization of TNP.

A novel cholesterol succinate – 1-Aminopyrene (CSA) Figure 1 structure (4) sensor for CL-20 (Hexanitro hexa azaiso wurtzitane) a nitramine explosive, was developed by Cheng Zhang et al. [21]. The synthesized CSA compound was demonstrated through UV-vis and florescence spectra, a characteristic aggregation –induced emission (AIE) effect in THF/H2O solution, which showed quenching the fluorescence in the presence of CL-20. The developed CSA sensor with AIE property was shown to be highly sensitive and specific for CL-20 in THF/H2O solutions with quenching constants of about 0.49×10^5, 1.29×10^5, 2.69×10^5 M respectively. CSA was shown to have high pH stability and high selectivity to CL-20. The fluorescence intensity could quench about 90% of the initial intensity, which could be identified by the naked eye under the UV light. The seal and cotton swabs treated with CSA were used for specific identification of CL-20. The synthesis of CSA is simple and cheap. The experimental work demonstrates that the CSA sensor is specific, selective, sensitive and fast for sensing the explosive CL-20 in practical security applications.

Kumbam Lingeshwar Reddy et al. [22] have reported the fluorescence detection of picric acid by using pyrene and anthracene based copper complexes. The authors have synthesized and characterized pyrene -copper complex (PCC) Figure 1 structure (5) as well as anthracene –copper complex (ACC). They evaluated the fluorescence behavior of both PCC and ACC with respect to nitro aromatics and certain anions. However in this presentation they have carried detailed work on the detection of the explosive, picric acid. Both the complexes were shown to be having sensitive fluorescence affinity with high selectivity towards picric acid at different pH ranges. The method is reported to be fast with naked eye observation, in practical applications in soil and other forensic fields.

Goodpaster J. V and McGuffin V. L [23] have carried out extensive work on the analysis of seventeen common nitrated explosives and their degradation products, which are electron acceptors, by fluorescence quenching of pyrene. They have determined hexahydro-1,3,5-trinitro-1,3,5-triazine (RDX), octahydro-1,3,5,7-tetranitro-1,3,5,7-tetrazine(HMX), 2,4,6-TNT, nitromethane, and ammonium nitrate in various commercial explosive samples by employing capillary liquid chromatography for separation, post column addition of pyrene solution and subsequent detection by indirect fluorescence quenching with laser induced fluorescence. This method shows increased sensitivity and selectivity over the UV-visible absorbance procedures and has the capability to detect wider range of organic and inorganic nitrated explosives. Nitro aromatic quenchers are shown to display unique wavelength dependence in their quenching constants as seen in residual structure after dividing unquenched and quenched emission spectra. This selective response is attributed to interactions of the nitro aromatic quenchers with excited-state pyrene molecules, which stabilize the excited state and shift the vibronic bands to slightly longer wavelengths. The application of the selective interaction between pyrene and nitro aromatic quenchers shows great promise in the development of novel sensors. Their studies further indicate that nitro aromatics have unique interactions with pyrene and the Stern-Volmer constants generally increase with the degree of nitration with aromatic explosives having more effective quenching effect than aliphatic or nitramine explosives. They suggest that sensors based on the pyrene quenching could be developed which would be more sensitive and selective when compared to other quenching compounds. The developed method is sensitive and discriminative for wide range of organic and inorganic explosives. The method could be used effectively in the sensors as well as a detection method in liquid chromatography. The procedure has applicability in environmental and forensic sciences.

Abu Saleh Musha Islam et al. [24] have designed and synthesized a new pyrene based fluorescent probe (N-(4- nitro-phenyl)-N'-pyren-1-ylmethyl-ene-ethane-1,2-diamine (PyDA-NP)) Figure 1 structure (6) capable of displaying dual emission; green fluorescence at ~ 523 nm for NO due to ICT process and aggregation -induced enhancement emission (AIEE) in the blue region at ~467 nm, for the formation of pyrene excimer complex. The mechanism of nitric oxide (NO/NO$^+$) sensing is based on N-nitrosation of aromatic secondary amine, which did not get interfered by reactive oxygen species and reactive nitrogen species. AIEE is attributed to the restriction of intra-molecular rotation and vibration, resulting in rigidity enhancement of the molecules. The AIEE property of the developed probe was characterized by UV−vis, fluorescence, DLS, SEM, TEM, XRD, optical fluorescence microscopy, and time-resolved photoluminescence studies. The probe gave maximum AIEE, with more than 800 fold increased fluorescence intensity with a high quantum yield of about 0.8, in a H_2O/CH_3CN (8:2 v/v) solution. In the aggregated state the probe could detect nitro explosives. They carried out work on the detection of 2, 4, 6, trinitro phenol (TNP) proving the charge transfer process of pyrene with electron deficient explosive molecules like TNP. The probe is claimed to be an excellent chemo sensor for selective detection of TNP in contact mode analysis. This non-dye based PyDA-NP molecule is a superior multifunctional material, which shows selective sensing of NO and selective detection of nitro explosive TNP.

Gupta S. K. et al. [25] designed and synthesized a Cu (I)-mediated pyrene-based fluorescent organic copolymer, with bulky isopropyl groups, Py-azo-COP Figure 1 structure (7). This polymer was investigated for its gas sensing and adsorption properties. The Py-azo-COP showed selective sensing of electron deficient poly nitro aromatic compound, picric acid. The surface area of the Py-azo-COP was measured to be 700 m^2 g^{-1} and it could store about 8.5 wt. % of CO_2 at 1 bar and 18.2 wt. % at 15.5 bar at 273 K. It also showed adsorption capability towards CO_2. The polymer showed hydrophobicity in adsorption studies with toluene and water. Due to the fluorescent nature of Py-azo- COP it could selectively sense picric acid among other poly nitro aromatic compounds such as DNT, p-DNB, and m-DNB, and other electron-deficient molecules. The authors suggest that further investigations could be carried out on the gas sorption properties and sensing capabilities of structurally similar COPs, specifically by removing the bulky isopropyl substituents for further increased porosity and possibly higher sensing capabilities. Work in this direction is stated to be in progress.

Fingerprints

Fingerprints, constituting a ridge pattern, are the most affirmative evidence of personal identification [26]. Individual leaves fingerprint on an object through pores that are present on each ridge which are attached to sweat glands in the skin. These ridges are arranged on the outer layer of skin, predominantly in the form of loops, whorls and arches (Figure 2) constituting the level 1 (primary) details. Level 2 (secondary) and level 3 (tertiary) details include minutiae and sweat pores respectively Figure 3 & 4. Minutiae constitutes ridge ending, bifurcation, dot, spur, island, bridge, enclosure, delta, double bifurcation etc. Generally, three types of fingerprints are encountered at crime scene: visible or patent, impression or indented and invisible or latent fingerprints [27]. Latent fingerprints are difficult to be directly detected by naked eyes, therefore, require development and enhancement techniques for their visualization. Traditional development methods include powder dusting, silver nitrate method, iodine fuming, superglue fuming, ninhydrin method etc. Powder dusting is primarily the preferred technique for the development of latent fingerprints which relies on the adherence of powder particles to the fingerprint residues, forming patterns of ridges and furrows for fingerprint development. The fingerprint powders are usually categorized as metallic, magnetic, colored or fluorescent. Among them, fluorescent powders are widely used on different surfaces for achieving high contrast under light irradiation. Fluorescent materials have significantly attracted the research interest due to their unique optical properties such as contrast, selectivity, sensitivity and minimum instrument requirement [28]. They are employed in the form of heavy metal quantum dots, noble metal nanoparticles, rare earth metal nanoparticles and fluorescent organic molecules for latent fingerprint development. Researchers recently are focused towards fluorescent organic molecules to develop environmental friendly and low toxic fingerprint developing materials unlike heavy metals powder which are potentially dangerous to the environment and users. Based on literature, one of the most studied organic chromophore is "Pyrene" [29]. Pyrene possesses high charge carrier mobility, π–π stacking behavior, exceptionally longer fluorescence lifetime and polarity-sensitive vibronic emission and chemical stability [30, 31].

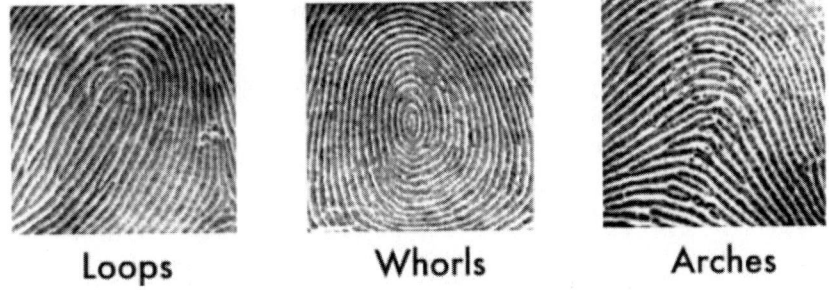

Figure 2. Different types of ridge patterns (Level 1 or primary).

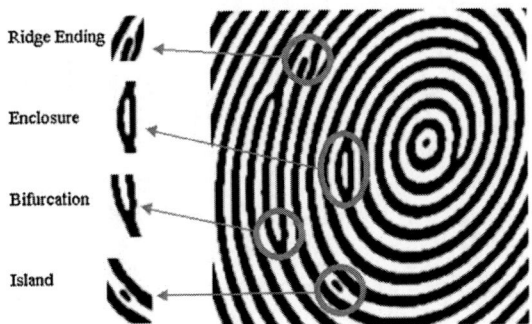

Figure 3. Minutiae - Level 2 (secondary) details in fingerprints.

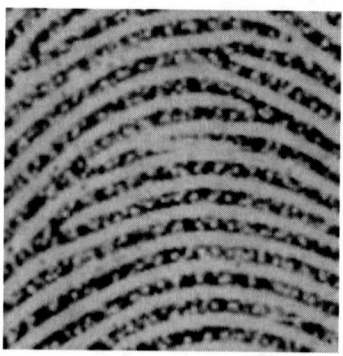

Figure 4. Sweat pores on fingerprint ridges - Level 3 (tertiary) details.

Considering pyrene properties, substituted imidazole tagged with fluorescent conjugated pyrene moieties namely 1-N-methyl-3-(2-oxo-2-(pyren-1-yl)ethyl)-imidazolium bromide (1), 1-N-isopropyl-3-(2-oxo-2-(pyren-1-yl)ethyl)-imidazolium bromide (2), 1-N-allyl-3-(2-oxo-2-(pyren-1-yl)ethyl)-imidazolium bromide (3), and 1-N-isopropyl-3-(2-oxo-2-(pyren-1-yl)ethyl)-imidazolium hexafluoro phosphate were designed, synthesized and characterized [32]. Optical properties of all the four compounds exhibited yellow and red color emissions in solid state based on the excitation wavelength which was examined by fluorescent microscope. The formulation of compound *1-N-isopropyl-3-(2-oxo-2-(pyren-1-yl)ethyl)-imidazolium hexafluorophosphate* Figure 6 structure (1) with neutral aluminum oxide G (TLC grade) showed individual characteristics of fingerprints with all the 3 level information (primary, secondary and tertiary characteristics) when viewed under fluorescence microscope. Depending on the excitation wavelength, fingerprints exhibit various color emissions as illustrated in Figure 5.

Enock et al. synthesized novel fluorescent labeled materials based on 3D polyhedral oligomeric silsesquioxanes (POSS) containing pyrene fluorophore Figure 6 structure (2) following the classical click reaction [33]. The CuI-catalyzed [3+2] cycloaddition between the POSS precursor with an azide group and the corresponding alkyne-like fluorophore was carried out to obtain the POSS-dye derivatives. The synthesized derivatives were characterized and were investigated for their photo physical properties. The photo physical study showed large Stoke's shift, good photo stability and displayed adequate physical characteristics to be successfully applied in latent fingerprint detection.

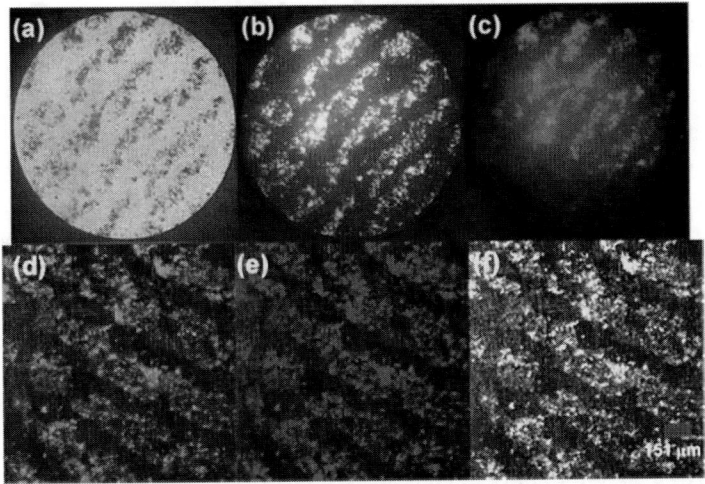

Figure 5. Fingerprints developed using compound *1-N-isopropyl-3-(2-oxo-2-(pyren-1imidazolium hexafluorophosphate* powder viewed through a confocal microscope on exposure to (a) bright field (b) blue and (c) green lights; fingerprints developed using compound 4-based fingerprint powder recorded using a confocal microscope on excitation of 488 nm laser and the emission monitored at (d) green and (e) red regions. (f) Overlaid image of d and e (scale bar = 151 μm) [32].

Sun et al. [34] constructed porous polymers linked by flexible cyclosiloxane units. 2,4,6,8-tetramethyl-2,4,6,8-tetravinylcyclotetrasiloxanes were reacted with brominated pyrene, tetraphenylethene, and spirobifluorene via the Heck reaction, resulting in three cyclosiloxane-linked fluorescent porous polymers Figure 6 structure (3). The synthesized materials exhibited strong fluorescence, tunable emission colors (with emission colors from brick-red at 606 nm to yellow at 549 nm and green at 489 nm while under UV light in the solid state and in suspensions) and high porosity. The materials showed effective fluorescence imaging of not only fresh LFPs but also aged LFPs with strong anti-interference ability in actual conditions.

Pyrene and its derivatives are fluorophores with strong intrinsic fluorescence, and are used as ligands to synthesize fluorescent metal-organic framework (MOF) materials [35]. Zr-based MOFs have been considered as excellent fluorescent probes for sensing explosives and ions [36]. Guo et al. used Zr as the metal center and 1,3,6,8-tetra (4-car- boxylphenyl) pyrene (TBAPy) and tetrakis(4-carboxyphenyl)por- phyrin (TCPP) as double linkers (ratio 5:1) to synthesize a novel Zr-MOF material Zr (TBAPy)$_5$(TCPP) via the solvothermal method with sequential linker installation [37]. For fingerprint detection using Zr-MOF, particle size plays an important role and can be adjusted using benzoic acid during the synthesis process [38]. Due to the fluorescence turn-on effect, it was found that Zr (TBAPy) 5 (TCPP) can efficiently achieve the identification of fingerprints.

Pyrene as such has the planar structure and has limited its use for development of electroluminescence materials due to π–π stacking and excimer formation in the solid state or in concentrated solution. However, pyrene when combined with rotor chromophores such as tetraphenyl ethylene (TPE) provides new fluorescent materials which emit light both in solution and in the aggregated state or solid state. Based on the strategy, D. Jana et al. [39] synthesized pyrene–vinyl–tetraphenylethylene based conjugated materials Figure 6 structure (4), (5), (6) and investigated the photo physical (including absorption, fluorescence, and fluorescence lifetime) and aggregation properties in tetrahydrofuran. Molar absorptivity for tetra substituted compounds was found to be higher when compared to mono substituted ones. The photo physical and aggregation behavior are dependent on structure, spacer and the number of methoxy groups. Based on the photo physical properties the molecule finds its application in biological sensors, latent fingerprint and explosive detection, chemo-sensors, solid state lighting etc.

P Karak et al. Discussed about cationic polycyclic hetero aromatic compounds (cPHACs) synthesis and their applications in cellular imaging and latent fingerprint visualization because of their high photo stability and biocompatibility [40]. Rapid synthesis of cPHACs was achieved by annulative alkyne-insertion π-extension (AAIPEX) in single step which proceeds via C-H activation of unfunctionalized heteroarene structure followed by alkyne insertion-annulation to give cPHACs.

Jaydev et al. [41] synthesized a new class of luminescent gold nanostructure by attaching the 1-pyreneiodide ligand to the surface of polyvinyl pyrrolidone-stabilized gold (Au: PVP) cluster. These Pyrene grafted nano clusters exhibited strong emission with the formation of an excimer within the neighboring pyrenes. Latent fingerprints were developed by dusting the powder of the Py-Au: PVP nano cluster over the fingerprints and viewed in day light and under 365 nm illumination. The fingerprints showed high clarity w.r.t. ridge pattern, ridge minutiae and pores on ridges. Authors also studied the development of aged fingerprints (30 days old) and found high degree of clarity. It is speculated that, for efficient chemisorption onto fingerprint residues, the combination of pyrene (hydrophobic nature) and PVP Au- cluster (hydrophilic nature) plays a significant role to make the powder sufficiently amphiphilic and adsorptive.

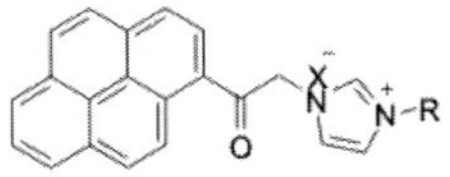

X = Br/ PF_6
R = Me/ iPr/ allyl

(1)

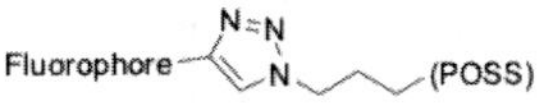

Fluorophore = pyrene, dansyl,
bromo-naphthalic anhydride, piperazine-naphthalic

(2)

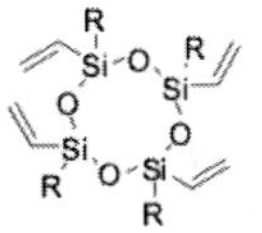

R = brominated pyrene,
tetraphenylethene, spirobifluorene

(3)

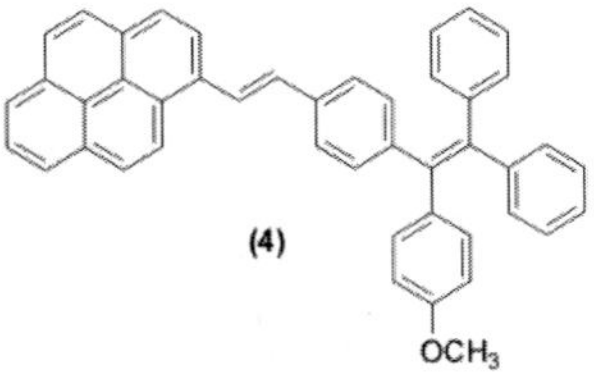

(4)

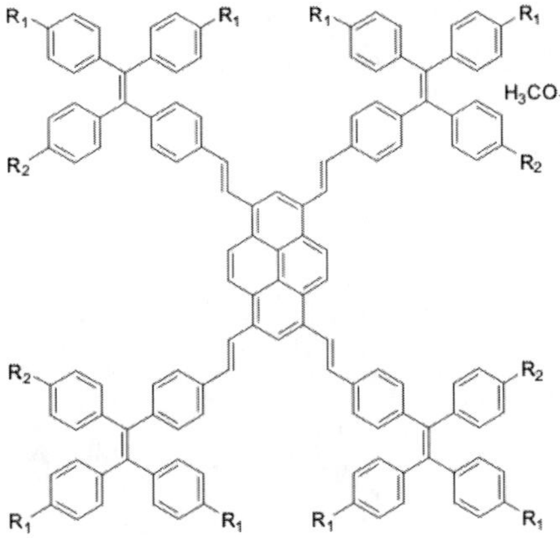

R_1 = H, R_2 = OCH_3
R_1 = OCH_3, R2 = H

(5)

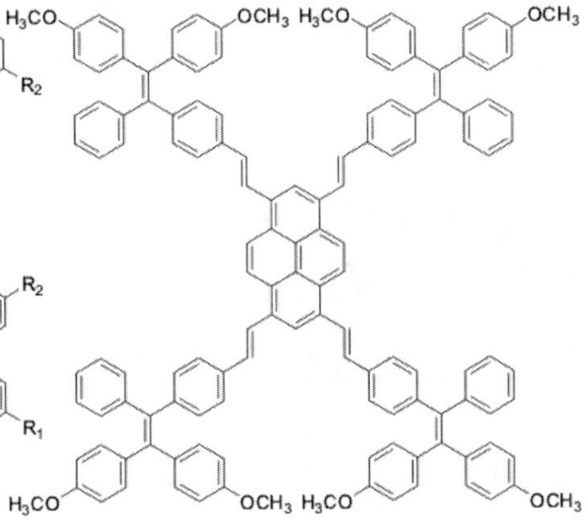

(6)

Figure 6. Novel molecules constituting pyrene moieties used for fingerprint detection.

Pyrene as such can also be used for the latent fingermark detection. Chang et al. conducted phase I studies using volatilized pyrene for visualizing latent fingermarks on nonporous surfaces [42]. Pyrene, when fumed, is absorbed onto the sebaceous components of fingermarks enabling their fluorescent visualization under UV excitation (Figure 7). The method was found effective on aluminum foil and glass and the pyrene fumed fingermarks retained fluorescence for many hours.

Our own study on fingermark detection using pyrene demonstrated the use of solid pyrene formulation with binders that can be efficiently used for the development of latent fingermarks on porous, non-porous and semi-porous surfaces [43]. Elaborate studies were conducted using varied pyrene concentrations followed by substrate study, time dependent study, temperature study, depleted fingermark development and the stability of the proposed formulation. When illuminated under 366 nm, the pyrene developed fingermarks showed clear, high contrast primary, secondary and tertiary level ridge details (Figure 8).

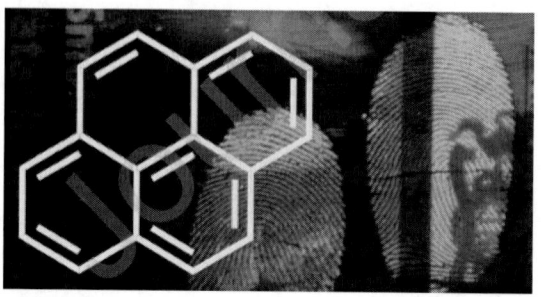

Figure 7. Pyrene fumed fingermarks [42].

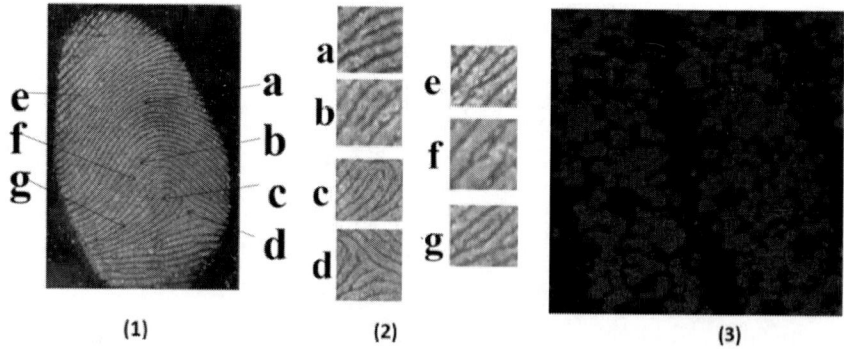

Figure 8. Fingermark showing (1) primary (2) secondary-minutiae (a) bifurcation (b) enclosure (c) core (d) delta (e) ridge end (f) short ridge/ island ridge (g) hook/ spur and (3) tertiary level details [43].

Bio Imaging

Pyrene and its derivatives have been extensively used as fluorescent probes due to its intermolecular excimer formation in solution and was first observed by Foster and Kasper in 1954 [44]. Three types of light emissions are present in pyrene fluorescent probes: Intramolecular charge transfer (ICT), Photo induced electron transfer (PET) and Forster resonance energy transfer (FRET).

The amalgamation of properties excimer formation, high quantum yield [45], long-lived excited states, detection of micro environmental changes [46], cell permeability and easy surface modification [47, 48] have made pyrene an unavoidable fluorescent organic molecule creating an impact in photochemical, photo physical and photo biological sciences.

In biology, Pyrene based sensors are used in a variety of applications. Pyrene probes are utilized in bio imaging and recognition of living cells, proteins, lipid membranes, peptides and DNA [49-54], polarity mapping of cells and embryos [55], phase separated biomolecules detection [56], glucose sensing using pyrene probes [57], pH and viscosity detection in lysosome [58].

J. Chao et al. [59] reported the synthesis of a novel pyrene fluorescent probe Figure 9 structure (1) for the detection of pH and HSO_3^- followed by their bio imaging in living cells. pH titrations showed a ratio metric emission (pH 7.10–1.36) with a pKa of 4.26, responded linearly to minor pH fluctuations within the acidic pH range of 3.00–5.50 in CH_3CN/H_2O (1/3, v/v), and exhibited a remarkable emission enhancement (pH 7.10–13.09) with a pKa of 10.91 and a linear response in the extremely alkaline pH range of 9.75–11.35 in CH_3CN/H_2O (1/19, v/v). The probe displayed a turn-off fluorescence signal in PBS (Phosphate buffered saline) buffer towards HSO_3^- which resulted in the conjugation breakage due to the Michael addition mechanism to the α,β-unsaturated ketone. The authors further extended the concept in intracellular fluorescent imaging of pH and HSO_3^- in living cells. The cellular studies showed that the probe, in biological system, was able to differentiate both the acidic and basic conditions and could detect the HSO_3^- in a cellular system. In yet another study, [60] the author J. Chao synthesized a pyrene based fluorescent probe Figure 9 structure (2) for cyanide detection. The probe comprises pyrene as electron donor and 2, 6-dichlorobenzene as electron acceptor linked by propylene ketone. The probe showed weakened fluorescence due to intramolecular charge transfer (ICT) [61, 62] phenomenon between the donor and the acceptor. In the presence of cyanide ion, nucleophilic attack occurs at the propylene ketone blocking the ICT process, resulting in increased fluorescence signal. The probe showed properties such as high selectivity and sensitivity with limit of detection at 37nM. After in vitro experimentation, the probe was successfully used for CN^- detection in cells, *C. Elegance* and Zebra fish. The probe was also employed to calculate the cyanide concentration in spike tap water samples as well.

Hossain et al. [63] reported a pyrene-appended bipyridine hydrazone-based ligand (HL) Figure 9 structure (3) that via self-assembly forms a hexa nuclear paddlewheel metal– organic macrocycle (MOM). When HL was reacted with Cu (II) ions it formed a fluorescent MOM with the molecular formula of [Cu_6L_6 (NO_3)$_6$]. The complex formed had 1:1 stoichiometry between HL and Cu (II) ions based on fluorescence titrations. In biological system, chemo-sensor HL showed high selectivity towards Cu(II) ions among other trace metals present with good cell permeability and low cytotoxicity (IC50~35mM).

Anupam et al. [64] reported a novel pyrene derivative bearing a benzilmonohydrazone moiety Figure 9 structure (4) for the selective sensing of Cu^{2+} ions. The molecule as a whole, exhibits weak fluorescence, due to PET process, from the nitrogen lone pairs to the pyrene moiety. In the presence of Cu^{2+} ions, the photo induced electron transfer mechanism is blocked resulting in enhanced fluorescent intensity. Visible color change was observed from light yellow to green in case of Cu^{2+} among other tested ions. The job's plot confirmed 1:1 stoichiometric ratio between pyrene derivative and Cu^{2+}. Bio imaging Cu^{2+} behavior was tested on human cervical cancer cells (HeLa). The cells showed no fluorescence when incubated with the Cu^{2+} and the pyrene derivative separately. However, green fluorescence was displayed after incubation with both pyrene derivative and Cu^{2+} indicating the derivatives' efficient use in imaging Cu^{2+} in cells.

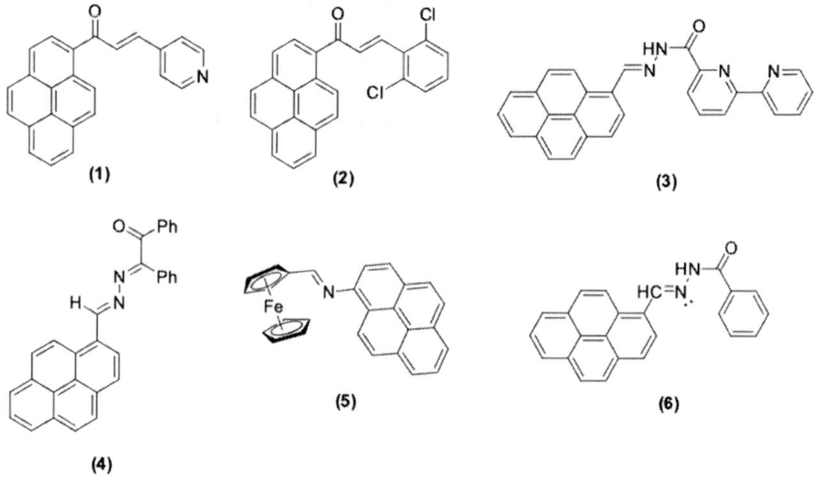

Figure 9. Novel molecules constituting pyrene moieties for bio imaging.

Mandeep Kaur et al. [65] reported a new dyad (1E)-N-(ferrocenyl-methylene)pyren-1- amine Figure 9 structure (5) that showed a *"turn-on"* chemodosimetric (analyte induced irreversible reaction) response in the presence of Cr^{3+} ion. The dyad as such was characterized with a weak emission band at 445 nm related to the monomer emission [66]. The fluorescence quantum yield ($\Phi = 0.016$) was low and showed quenched pyrene emission due to photo induced electron transfer (PET) from the ferrocenyl–imine unit to the excited pyrenyl subunit [67]. In this case, the analyte (Cr^{3+}) triggered hydrolysis reaction broke the imine functionality leading to the formation of 1-aminopyrene, a "*turn-on*" response. Cell imaging application for the detection of Cr^{3+} in MCF-7 cells (breast cancer cells), the dyad showed fluorescence in perinuclear region of the cells, indicating its subcellular distribution and a very good membrane permeability.

Saravanan et al. [68] for the first time reported the use of pyrene based Schiff base derivative (nicotinic acid pyren-1-ylmethylene-hydrazide) Figure 9 structure (6) for Bi^{3+} ion detection (DMSO-H2O, 1:1 v/v, HEPES = 50 mM, pH = 7.4). It showed dual "turn-on" response for the detection of Bi^{3+} and Al^{3+} ions. The derivative, a yellow crystalline solid was synthesized using 1-pyrenecarboxaldehyde and nicotinic hydrazide dissolved in ethanol followed by 8hrs reflux. Due to the photo-induced electron transfer (PET) process between azomethine (C=N) nitrogen lone pair and the pyrene moiety, the derivative exhibited weak fluorescence. However, upon addition of the Bi^{3+} and Al^{3+} ions to the solution containing the derivative, the PET process was forbidden leading to the enhanced fluorescent intensity. Bi^{3+} exhibited bright green whereas light green was observed in case of Al^{3+}. Fluorescence titration was performed by the gradual addition of Bi^{3+} and Al^{3+} ions (0 to 100 equiv.) to the nicotinic acid pyren-1-ylmethylene-hydrazide solution and showed prominent fluorescent intensity for the selected ions even in the presence of other coexisted metal ions. The derivative as such exhibited low quantum yield ($\Phi = 0.089$) however, upon complexation with Bi^{3+} and Al^{3+}, the enhanced quantum yield value was $\Phi = 0.681$ and 0.575 respectively. Both the ions took nearly 3-4 minutes for the process of completion of binding with the derivative. Additional reversibility binding studies using Na_2EDTA titration experiments showed decrease in fluorescence intensity reaching almost the original fluorescence intensity due to EDTA-Bi^{3+} and EDTA-Al^{3+} formation. The synthesized derivative was further used for the fluorescent bioimaging of Bi^{3+} and Al^{3+} in RAW 264.7 cells. The observations include minimal cytotoxicity (IC50 ~94μM), bio sensing both the Al^{3+} and Bi^{3+} ions with superior fluorescent intensity towards the later and good cell permeability.

Jablonski et al. [69] reported the synthesis of Pyrene-nucleobase conjugates namely pyrene-adenine and pyrene-thymine for DNA recognition and bio imaging applications. The pyrenyl group and the nucleobases are known to act as intercalator and to selfassemble via hydrogen bond respectively. The conjugates with hydroxy functional groups displayed high emission quantum yield (40%) and longer fluorescence decay times (~150 ns) compared to the carbonyl conjugates. Interaction with oligonucleotides was investigated both in water and in buffer solution. The adenine conjugates (the carbonyl and the hydroxy) self-assembled on oligomer in water solution as per canonical base-base pairing. However, in buffered solution, only the hydroxy conjugate could bind effectively to the oligomer. Besides, hydroxy conjugate of pyrene-adenine assembly showed self-assembly ratio of 112% to the double-stranded oligonucleotide which may be due to triple-helix-like binding, intercalation, or both.

Cations and Anions

Certain cations and anions are forensically important. These include toxic metals involved in poisoning such as Zn, Hg, Pb, Fe, Cu, Ba, Sb, and As. Certain metals found in the gun shot residues such as Ni, Al, Fe, Sn, Cu, Sr, Zn and Ti are relevant. In poisoning cases sometimes certain toxic anions are also required to be determined such as cyanide, phosphide, nitrates, sulfates, phosphates, chromate, fluoride etc. In gunshot residues and explosion residues certain anions are required to be determined apart from the above cations. Pyrene and its derivatives are used as sensors for the detection of metal ions in the environmental and other samples. It is reported that sensing of anions is quite difficult and therefore it is considered as a fertile area of research. In this review we have covered fluorescent sensors based on pyrene and its derivatives for determination of metal cations. In the area of anion sensing in the forensic field we did not find many papers except for fluoride. The literature on the forensic application of the pyrene based sensors is almost negligible and hence this area can be explored in forensic work. Most of the work is reported in water, waste water and in tissues. Kannan A. and Praveena V. have published an excellent review in 2021 on Pyrene based materials as fluorescent probes for cations, anions and neutral organic molecules [70].

Cations: A fluorescent sensor with pyrene and thiophene Figure 10 structure (1) was synthesized by Wu, et al. in 2015. This compound is used to detect Cu^{2+} in the presence of other metal ions based on Forster Resonance Energy Transfer (FRET) mechanism. The monomer–excimer conversion takes place when the pyrene-thiophene compound is induced by Cu^{2+}, leading to the formation of a new broad band at 460 nm, corresponding to the pyrene excimer [71]. Yin-Shuang Wu et al. prepared a pyrene based fluorescent sensor for copper ions in living cells. A simple pyrene-thiophene compound was synthesized which acts a ratiometric chemo sensor for Cu^{2+}. For this a simple pyrene – thiophene compound was synthesized which is used to sense Cu^{2+} ions. The Cu^{2+} ions induce the monomer- excimer conversion ratiometric signal. The chemo sensor shows a linear range response for Cu^{2+} from 1.0×10^{-7} to 2.0×10^{-5} M with the detection limit of 2×10^{-8} M. The synthesized compound shows good response for Cu^{2+} over a pH range of 3.0 – 8.0. The chemo sensor is specific for Cu^{2+} as other metal ions do not interfere. This chemo sensor response is fast for Cu^{2+}. The authors have demonstrated the application of this sensor for determining Cu^{2+} in water and image Cu^{2+} in living cells.

A pyrene based Schiff base derivative having AIE phenomena and a fluorescent turn-on sensor property was reported by Wu et al. in 2017. A Schiff base derivative of pyrene Figure 10 structure (2) was prepared by Wu *et al.* by one – pot synthesis, which was developed as a fluorescent turn-on sensor for Cu^{2+} ions. Due to paramagnetic nature of Cu^{2+} the probe shows a quenching effect. This compound is synthesized by a one-pot multi-component reaction. The Schiff base derivative by virtue of AIE phenomenon, is highly selective for Cu^{2+} ions in HeLa cells in situ [72].

D. Rajasekaran et al. synthesized in 2019 a pyrene derivative [(3-((1,8-dihydropyrene-1-yl)methylene)pentane-2,4-diylidene) bis (hydrazine-1-yl-2-ylidene bis benzothiazole] (DPD) Figure 10 structure (3) for detecting Cu^{2+} in live-cell imaging and in water samples. It was found to be highly selective to Cu^{2+} ions in the presence of other metal ions such as K^+, Na^+, Zn^{2+}, Ni^{2+}, Pb^{2+}, Cd^{2+}, Mg^{2+}, Hg^{2+}, Ag^+, Mn^{2+}, Fe^{3+} and Al^{3+} under the physiological pH conditions [73].

Chakraborty et al. [74] synthesized an ideal pyrene based compound as a sensor in 2018 for sensing copper ions. The compound was pyrene and 4-aminoantipyrine linked via a Schiff-base Figure 10 structure (4). In this sensor, the pyrazolone moiety serves for chelation purposes as also for radical scavenging. In a water–acetonitrile mixture, the compound shows an aggregation induced emission (AIE) property. The critical aggregation concentration (CSC) was found to be 23.4 mM. The sensor exhibits a selective turn on emission showing visible colorimetric changes. The compound also showed antioxidant activity.

Phapale et al. [75] reported in 2017 a simple and cheap pyrene based chemo sensor Figure 10 structure (5) for sensing Cu^{2+} and Fe^{3+} ions in biological and chemical samples. 1: 1 and 2: 1 stoichiometric binding of Cu^{2+}

and Fe^{3+}, respectively, was demonstrated through Job plots and ESI-MS analysis. A 660 nm band in the UV-vis absorption region and the strong color shift visible to the naked eye indicated the presence of Cu^{2+} ions and suggesting that Cu^{2+} ions are more selectively bound than other metal cations.

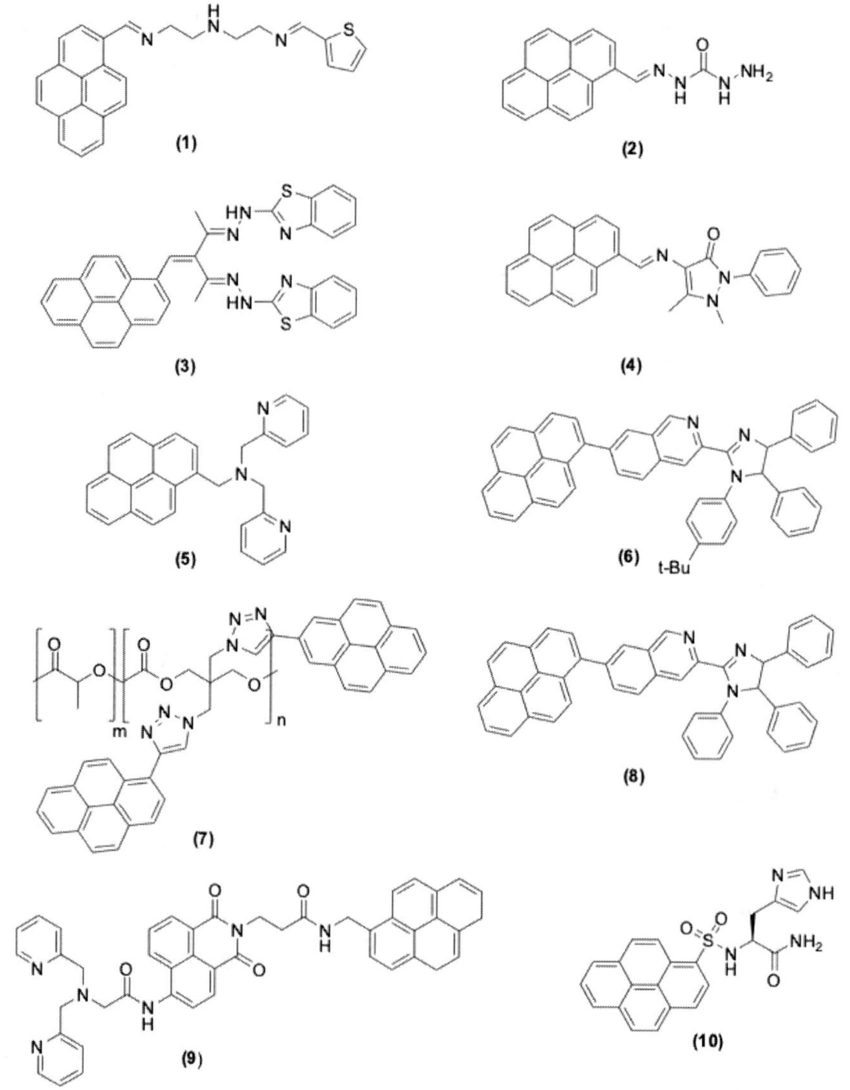

Figure 10. Novel molecules constituting pyrene moieties for ions detection.

Harsha et al. [76] synthesized in 2020, 2-yl]-6-(pyren-1-yl) quinoline (TDIPQ) Figure 10 structure (6) as an ON-OFF fluorescent chemo sensor for copper(II) ions. Colorless TDIPQ in acetonitrile–water (2:1, v/v) selectively turns yellow along with fluorescence quenching upon addition of copper (II) ions. The fluorescence quenching was reported to be directly proportional to the concentration of copper (II) ions. The interaction between TDIPQ and copper (II) was studied with the help of UV-Vis, fluorescence, ^{1}H NMR, and MALDI mass spectral techniques. The Job's plot gave the stoichiometry of the TDIPQ–Cu complex as 2:1. The other metal ions showed negligible or no interference at all in the detection ability of the sensing compound. The detection limit of TDIPQ was shown to be 2×10^{-6} M and the quantitation limit to be 6.2×10^{-6} M. The reported method had a precision of 0.98 ± 0.011 and recovery $99.09 \pm 1.4\%$. The method was shown to be applicable for determining Copper (II) in variety of matrices such as potable water, waste water, and soil. Further a mixture of TDIPQ with the BZA-Co-BZMA polymer can be used as a film on glass as a sensor to indicate the presence of copper, which was assessed by SEM imaging.

A novel polylactide (PLA) polymer containing pyrene side groups Figure 10 structure (7) was synthesized in 2021 by Erdinc Doganci [77] which could be used as a chemo sensor for nitro aromatic compound (NAC) and heavy metal ion (HMI). The novel polymer was prepared in a four-step reaction sequence, including azidification, esterification, ring opening polymerization and click reactions. First, an azido functional cyclic carbonate monomer (ADTC) was synthesized and then a copolymer of l-lactide and ADTC monomers was synthesized to obtain PLA with azido functional groups (P (LA-co-ADTC)). Then the compound was modified with pyrene groups using 'click chemistry ' which is a highly effective modification method. The CH molecular structures and other properties were determined by FTIR, GPC and ^{1}H NMR. The fluorescence characteristics were studied by UV–visible and fluorescence spectrometry. The polymers were used as a fluorescent probe to determine Zn^{2+}, Hg^{2+}, Mn^{2+}, Pb^{2+}, Cd^{2+}, Co^{2+} compounds and NACs (1,2-dinitrobenzene, 2,4-dinitrotoluene, 2,4,6-trinitrophenol (picric acid, PA), 4-nitrophenol (4-NP), 2,4-dinitrophenol (2,4-DNP), 2,4,6-trinitrotoluene, 4-nitrotoluene). The fluorescence intensity of the polymers decreased linearly with the addition of NACs and HMIs. The simplicity and sensitive nature of the quenching property of the polymers was used effectively to determine metal ions as well as nitro aromatic compounds.

In 2018, Harsha et al. [78] synthesized a novel pyrene tethered imidazole derivative 6-(Pyren-1-yl)-2-(1,4,5-triphenyl-1H-imidazol-2-yl) quinoline

(PTIQ) Figure 10 structure (8) which could be used as a selective sensor for Hg^{2+} in various matrices. PTIQ is colorless in CH_3CN buffer (2:1 v/v, pH 5.0) which turns to yellow with quenching its fluorescence in the presence of Hg^{2+} ions. The chemistry of PTIQ with Hg^{2+} was studied using NMR and MALDI-MS. The MDL and LOQ were found to be $9.8 \times 10^{-8}M$ and $2.95 \times 10^{-7}M$, respectively. The method claims excellent precision and recovery. The proposed method claims negligible interference from other metals and good sensitivity with PTIQ. The authors have applied the proposed method for determining Hg^{2+} in diverse samples such as Millipore water, soil, laboratory waste water, blood, urine, NaCl and dump yard waste, the values of which are comparable with the established methods.

Walekar, Laxman S. et al. [79] prepared pyrene-AgNPs system for detection of Hg (II) based on Forster Resonance Energy Transfer (FRET) principle. The integrated system of pyrene and cetyltrimethyl ammonium bromide (CTAB) capped silver nanoparticles (AgNPs) with a distance of 2.78 nm were developed for the detection of Hg (II) and pyrene dimer. The interaction between pyrene and AgNPs results in the fluorescence quenching of pyrene due to the energy transfer, whose mechanism can be attributed to the FRET. The probe prepared by the authors shows a highly selective and sensitive response towards Hg (II) with a 90% fluorescence recovery and change of color from yellowish brown to colorless. The authors claim selective response from Hg (II) and non-interference of other metal ions. The proposed method has linearity between 0.1 and 0.6 µg mL^{-1} with a detection limit of 62 ng mL^{-1}. The system also claims it as an effective method for detection of pyrene in its dimer form even at very low concentrations (10 ng mL^{-1}) on the surface of AgNPs. The proposed method could be used for the detection of Hg (II) as well as pyrene simultaneously.

Pyrene was attached with naphthalimide–dipicolylamine to prepare a pyrene derivative Figure 10 structure (9) by Yoon et al. [80] in 2017 for the sensing of Zn^{2+} ions. The derivative is able to detect Zn^{2+} ion fluorescence intensity for Zn^{2+} ions, and the naphthalimide–DPA moiety works as a Zn^{2+} sensing device giving the fluorescence shift. The authors have used the proposed method for determining Zn^{2+} ions in a drug supplement. They suggest that fluorescent color changes from blue to green take place due to Photo induced Electron Transfer (PET) in the presence of Zn^{2+} ions. They observed that the determination of Zn^{2+} is viable in a wide range of pH from 4 – 11.

A pyrene fluorescent chemo sensor was developed by Thirupathi and Lee [81] by attaching the histidine to Figure 10 structure (10) pyrene which gave a sensing response towards Zn^{2+} in the pH range of 7.5 to 11.5. The addition of Ag^+ and Cu^{2+} quenched the fluorescence emission. The derivative is highly selective and sensitive to Zn^{2+} ions among the tested metal ions. When the derivative reacts with Zn^{2+}, the resulting complex showed fluorescence enhancement at 485 nm due to the formation of excimer resulting in the fluorescent color change from blue to bluish-green. The compound when it reacts with Zn^{2+} in the presence of H_2PO_4 and cysteine was able to retrieve the fluorescence. This derivative can be used, as suggested by the authors for determining Zn^{2+} and two other bio-chemicals.

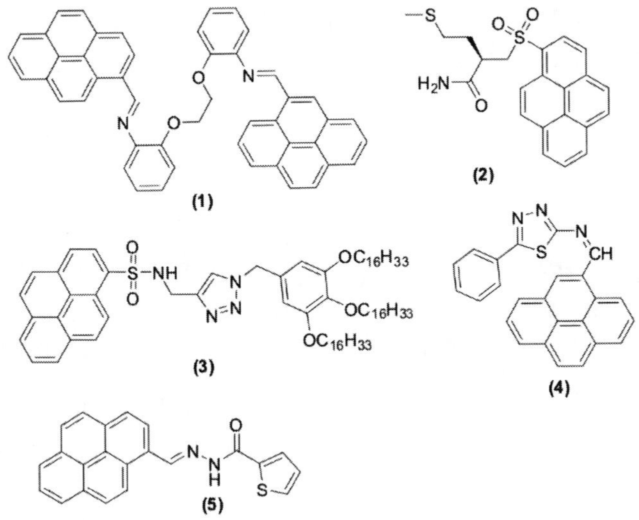

Figure 11. Novel molecules constituting pyrene moieties for ions detection.

In 2010, Zhou et al. [82] prepared a selective ratiometric fluorescent probe by combining pyrene with azadiene Figure 11 structure (1) group for sensing Hg^{2+} ions. The detection is based on the monomer–excimer emission due to the conformational changes upon interaction of the derivative with Hg^{2+} ions. The authors report that the synthesis of the derivative is simple and it is stated to be stable in both alkaline and acid solutions.

A probe consisting of pyrene derivative and an amino acid (Py-Met) Figure 11 structure (2) was prepared by Yang et al. [83] in 2011 to detect Hg^{2+} in environmental samples. The synthesis of the derivative is based on a sulfonamide-induced metal ion deprotonation mechanism and it shows selective and sensitive fluorescence with a decrease in the monomer upon interaction with Hg^{2+} ions in the tested samples. Amino acid-based fluorescence sensors are stated to have the ability to detect metal ions in aqueous solutions such as environmental samples due to their solubility in water.

By utilizing the click reaction Liu et al. [84] reported a new compound Figure 11 structure (3) in 2018, wherein they prepared a amphiphilic probe by reacting pyrene with a cetyl chain. The pyrene acts as a fluorophore in the hydrophobic cetyl chain. This compound was tested in DMF solution, aqueous solution and gel state to determine Hg^{2+}. The derivative showed pyrene-based monomer emission in DMF solution binding to Hg^{2+} in a 1 : 1 stoichiometry with the fluorescence serving as a turn-off signal. In aqueous solution, the probe shows self-assemblance of nano aggregates and gives both pyrene-based monomer and excimer emissions. As Hg^{2+} is bound to the generated nano aggregates, the monomer emission is mostly quenched, but the excimer emission changes negligibly. The probe in aqueous solution was found to be more responsive to Hg^{2+} than in DMF solution. The probe develops fluorescent organic gels in certain organic solvents, which collapsed with the addition of Hg^{2+} ions. The authors have successfully applied the developed probe for determining intracellular Hg^{2+} ions by fluorescence imaging in biomedical applications.

In 2019 Bai et al. [85] synthesized a novel long-wavelength turn-on fluorescent probe based on pyrene to identify Hg^{2+} Figure 11 structure (4). The derivative was found to sense Hg^{2+} in the presence of other metal ions. It showed turn-on fluorescence emission at 607 nm with a red shift in the absorption spectrum. The transition of color of the solution from yellow to orange could be seen by the naked eye. When it was added to a solution of the derivative and Hg^{2+}, the fluorescence of the probe was found to be reversible. Based on the Job's plot, electrospray ionization mass spectrometry (ESI-MS), SEM, and DFT, the relationship between the probe and Hg^{2+} was seen. The derivative could detect Hg^{2+} using the ''off–on–off ''fluorescent signal. The derivative showed a high selectivity for Hg^{2+}. The authors suggest that the probe can also be used to detect Hg^{2+} on test strips and silica gel plates.

A fluorescent chemo sensor based on homooxacalix-[3]arene was reported by Ni et al. [86] in 2011, showing a high affinity towards Pb^{2+} ions. This chemo sensor when attached with pyrene forms two compounds which were reported concurrently. The selective and sensitive detection of Pb^{2+} ions is due to the homooxacalix- [3]arene which avoids fluorescence quenching incorporated with pyrene which provides an enhanced monomer emission in an aqueous organic solution.

A fast and cost-effective AIE probe with a special switchable color shift in the solution for F ions was reported in 2019 by Yadav et al. [87] The compound Figure 11 structure (5) is a pyrene based fluorescent probe, named (E)-N-(pyren-1- yl methylene)thiophene-2-carbohydrazide (PTH), with high selectivity and sensitivity towards F ions compared to several other anions. A distinct color change upon interaction with fluoride ions which can be seen with naked eye in a ACN/water medium as a strong bright yellow emission. The pyrene fluorophore in the compound endures a non-radiative decay owing to a π–π stacking interaction and it functions as an ACQ luminophore between planar pyrene rings. It also showed AIE enhancement in the binary solvent.

Conclusion

It is said that pyrene is the fruit fly of photo-chemists. Its unique properties have inspired researchers from many scientific areas, including forensic science, making pyrene the chromophore of choice in fundamental and applied photochemical research.

Pyrene and its derivatives are important polycyclic aromatic hydrocarbons that act as fluorophores and are widely used as a fluorescent probe on account of their strong emission of fluorescence in live cells, very low cytotoxicity, high fluorescence quantum yield, easy modification, and outstanding cell permeability. Its derivatives are also valuable molecular probes via fluorescence spectroscopy, having a high quantum yield and lifetime. It has strong absorbance in UV-Vis in three sharp bands at 330 nm in DCM.

Quenching of the monomer and excimer emissions from pyrene allows excellent discrimination in the detection of electron-deficient molecules; this characteristic could be used to detect explosives in various matrices. A novel polylactide polymer containing pyrene side groups have been used as a fluorescent chemical probe towards certain heavy metal ions and nitro-aromatic compounds.

While this chapter deals with the applications of pyrene derivatives for the determination of explosives, nitro-aromatics, cations, anions, fingerprints and bio-imaging in forensic science perspective.

It is felt that there could be more potential opportunities in analytical forensics based on immense possibilities of synthesizing new pyrene derivatives to be used as chromophores and probes.

References

[1] Worawong CH, Phutdhawong WA, Jirasirisak S. The study of fluorescent chemicals for fingerprint development. *In 2016 9th Biomedical Engineering International Conference* (BMEiCON) 2016 Dec 7 (pp. 1-3). *IEEE*.

[2] Thornton JI. Modification of fingerprint powder with coumarin 6 laser dye. *Journal of Forensic Science*. 1978 Jul 1; 23(3): 536-8.

[3] Almog J, Gabay A. Chemical reagents for the development of latent fingerprints. III: Visualization of latent fingerprints by fluorescent reagents in vapor phase. *Journal of Forensic Science*. 1980 Apr 1; 25(2):408-10.

[4] Sodhi GS, Kaur J, Garg RK, Kobilinsky L. A fingerprint powder formulation based on rhodamine B dye. *Journal of Forensic Identification*. 2003 Sep 1; 53(5): 551.

[5] Liu L. Study on the use of rhodamine doped nanocomposite for latent fingerprint detection. In *Advanced Materials Research* 2011 (Vol. 295, pp. 813-816). Trans Tech Publications Ltd.

[6] Sharma KK, Nagaraju P, Mohanty ME, Baggi T R and Rao VJ 2018 Latent fingermark development using a novel phenanthro imidazole derivative *J. Photochem. Photobiol.* 51, 253–60.

[7] Wang M, Zhu Y, Mao C. Synthesis of NIR-responsive NaYF4: Yb, Er upconversion fluorescent nanoparticles using an optimized solvothermal method and their applications in enhanced development of latent fingerprints on various smooth substrates. *Langmuir*. 2015 Jun 30; 31(25):7084-90.

[8] Tang HW, Lu W, Che CM, Ng KM. Gold nanoparticles and imaging mass spectrometry: double imaging of latent fingerprints. *Analytical chemistry*. 2010 Mar 1; 82(5):1589-93.

[9] Choi MJ, McBean KE, Ng PH, McDonagh AM, Maynard PJ, Lennard C, Roux C. An evaluation of nanostructured zinc oxide as a fluorescent powder for fingerprint detection. *Journal of Materials Science*. 2008 Jan; 43(2):732-7.

[10] Theaker BJ, Hudson KE, Rowell FJ. Doped hydrophobic silica nano-and micro-particles as novel agents for developing latent fingerprints. *Forensic Science International*. 2008 Jan 15; 174(1):26-34.

[11] Wolfbeis OS. Nanoparticle-enhanced fluorescence imaging of latent fingerprints reveals drug abuse. *Angewandte Chemie International Edition*. 2009 Mar 16; 48(13):2268-9.

[12] Herkstroeter WG, Martic PA, Hartman SE, Williams JL, Farid S. Unique hydrophobic interactions of pyrene in aqueous solution as effected by polyelectrolytes and surfactants. *Journal of Polymer Science: Polymer Chemistry Edition*. 1983 Aug; 21(8):2473-90.

[13] Niko Y, Cho Y, Kawauchi S, Konishi GI. Pyrene-based D–π–A dyes that exhibit solvatochromism and high fluorescence brightness in apolar solvents and water. *RSC Advances*. 2014; 4(69): 36480-4.

[14] Singla P, Kaur P, Singh K. Discrimination in excimer emission quenching of pyrene by nitroaromatics. *Tetrahedron Letters*. 2015 Apr 29; 56(18):2311-4.

[15] Bai H, Li C, Shi G. Rapid nitroaromatic compounds sensing based on oligopyrene. *Sensors and Actuators B: Chemical*. 2008 Mar 28; 130(2):777-82.

[16] Boonsri M, Vongnam K, Namuangruk S, Sukwattanasinitt M, Rashatasakhon P. Pyrenyl benzimidazole-isoquinolinones: Aggregation-induced emission enhancement property and application as TNT fluorescent sensor. *Sensors and Actuators B: Chemical*. 2017 Sep 1; 248:665-72.

[17] Focsaneanu KS, Scaiano JC. Potential analytical applications of differential fluorescence quenching: pyrene monomer and excimer emissions as sensors for electron deficient molecules. *Photochemical & Photobiological Sciences*. 2005 Oct; 4(10): 817-21.

[18] Thakarda J, Agrawal B, Anil D, Jana A, Maity P. Detection of trace-level nitroaromatic explosives by 1-pyreneiodide-ligated luminescent gold nanostructures and their forensic applications. *Langmuir*. 2020 Dec 8; 36 (50): 15442-9.

[19] Taudte RV, Beavis A, Wilson-Wilde L, Roux C, Doble P, Blanes L. A portable explosive detector based on fluorescence quenching of pyrene deposited on coloured wax-printed μPADs. *Lab on a Chip*. 2013;13(21):4164-72.

[20] Shyamal M, Maity S, Mazumdar P, Sahoo GP, Maity R, Misra A. Synthesis of an efficient Pyrene based AIE active functional material for selective sensing of 2, 4, 6-trinitrophenol. *Journal of Photochemistry and Photobiology A: Chemistry*. 2017 Jun 1; 342:1-4.

[21] Zhang C, Pan X, Cheng S, Xie A, Dong W. Pyrene Derived aggregation-induced emission sensor for highly selective detection of explosive CL-20. *Journal of Luminescence*. 2021 May 1; 233:117871.

[22] Reddy KL, Kumar AM, Dhir A, Krishnan V. Selective and sensitive fluorescent detection of picric acid by new pyrene and anthracene based copper complexes. *Journal of fluorescence*. 2016 Nov; 26(6):2041-6.

[23] Goodpaster JV, McGuffin VL. Fluorescence quenching as an indirect detection method for nitrated explosives. *Analytical Chemistry*. 2001 May 1;73 (9):2004-11.

[24] Islam AS, Sasmal M, Maiti D, Dutta A, Show B, Ali M. Design of a pyrene scaffold multifunctional material: real-time turn-on chemosensor for nitric oxide, AIEE behavior, and detection of TNP explosive. *ACS omega*. 2018 Aug 31;3 (8):10306-16.

[25] Gupta SK, Kaleeswaran D, Nandi S, Vaidhyanathan R, Murugavel R. Bulky isopropyl group loaded tetraaryl pyrene based azo-linked covalent organic polymer for nitroaromatics sensing and CO2 adsorption. *ACS omega*. 2017 Jul 13;2 (7):3572-82.

[26] Hazarika P, Russell DA. Advances in fingerprint analysis. *Angewandte Chemie International Edition*. 2012 Apr 10;51(15):3524-31.

[27] Datta AK, Lee HC, Ramotowski R, Gaensslen RE. *Advances in fingerprint technology*. CRC press; 2001 Jun 15.

[28] Lian J, Meng F, Wang W, Zhang Z. Recent trends in fluorescent organic materials for latent fingerprint imaging. *Frontiers in Chemistry*. 2020 Nov 9; 8:594864.

[29] Kobayashi H, Ogawa M, Alford R, Choyke PL, Urano Y. New strategies for fluorescent probe design in medical diagnostic imaging. *Chemical reviews*. 2010 May 12; 110 (5):2620-40.

[30] Winnik FM. Photophysics of preassociated pyrenes in aqueous polymer solutions and in other organized media. *Chemical reviews*. 1993 Mar 1; 93 (2):587-614.

[31] Berlman I. *Handbook of florescence spectra of aromatic molecules*. Elsevier; 2012 Dec 2.

[32] Nirmala M, Vadivel R, Chellappan S, Malecki JG, Ramamurthy P. Water-Soluble Pyrene-Adorned Imidazolium Salts with Multicolor Solid-State Fluorescence: Synthesis, Structure, Photophysical Properties, and Application on the Detection of Latent Fingerprints. *ACS omega*. 2021 Apr 9; 6 (15):10318-32.

[33] Dare EO, Vendrell-Criado V, Consuelo Jimenez M, Pérez-Ruiz R, Diaz Diaz D. Fluorescent-Labeled Octasilsesquioxane Nanohybrids as Potential Materials for Latent Fingerprinting Detection. *Chemistry–A European Journal*. 2020 Oct 15; 26 (58):13142-6.

[34] Sun R, Feng S, Zhou B, Chen Z, Wang D, Liu H. Flexible cyclosiloxane-linked fluorescent porous polymers for multifunctional chemical sensors. *ACS Macro Letters*. 2019 Dec 19;9 (1):43-8.

[35] Guo L, Wang M, Cao D. A novel Zr-MOF as fluorescence turn-on probe for real-time detecting H2S gas and fingerprint identification. *Small*. 2018 Apr;14 (17):1703822.

[36] Nagarkar SS, Desai AV, Ghosh SK. A fluorescent metal–organic framework for highly selective detection of nitro explosives in the aqueous phase. *Chemical Communications*. 2014; 50 (64):8915-8.

[37] Guo L, Zeng X, Lan J, Yun J, Cao D. Absorption competition quenching mechanism of porous covalent organic polymer as luminescent sensor for selective sensing Fe^{3+}. *Chemistry Select*. 2017 Jan 23;2(3):1041-7.

[38] Bai Y, Dou Y, Xie LH, Rutledge W, Li JR, Zhou HC. Zr-based metal–organic frameworks: design, synthesis, structure, and applications. *Chemical Society Reviews*. 2016 45 (8):2327-67.

[39] Jana D, Boxi S, Parui PP, Ghorai BK. Planar–rotor architecture based pyrene–vinyl–tetraphenylethylene conjugated systems: photophysical properties and aggregation behavior. *Organic & Biomolecular Chemistry*. 2015; 13 (43):10663-74.

[40] Karak P, Rana SS, Choudhury J. Cationic π-extended heteroaromatics via a catalytic C–H activation annulative alkyne-insertion sequence. *Chemical Communications*. 2022; 58 (2):133-54.

[41] Thakarda J, Agrawal B, Anil D, Jana A, Maity P. Detection of trace-level nitroaromatic explosives by 1-pyreneiodide-ligated luminescent gold nanostructures and their forensic applications. *Langmuir*. 2020 Dec 8; 36 (50):15442-9.

[42] Chang I, Stone AC, Hanney OC, Gee WJ. Volatilised pyrene: A phase 1 study demonstrating a new method of visualising fingermarks with comparisons to iodine fuming. *Forensic science international*. 2019 Dec 1; 305:109996.

[43] Sharma KK, Kannikanti GH, Baggi TR, Vaidya JR. A pyrene formulation for fluorometric visualization of latent fingermarks. *Methods and Applications in Fluorescence*. 2018 Apr 25; 6 (3):035004.

[44] Forster T, Kasper K. Physik Chem (Frankfurt). *J Phys Chem*. 1954;1:275-82.

[45] v. Bünau G. *JB Birks: Photophysics of Aromatic Molecules*. Wiley-Interscience, London 1970. 704 Seiten. Preis: 210s.

[46] Kalyanasundaram K, Thomas JK. Environmental effects on vibronic band intensities in pyrene monomer fluorescence and their application in studies of

micellar systems. *Journal of the American Chemical Society*. 1977 Mar; 99 (7):2039-44.

[47] Meech SR, Phillips D. Photophysics of some common fluorescence standards. *Journal of photochemistry*. 1983 Jan 1; 23 (2):193-217.

[48] Kumar A, Pandith A, Kim HS. Pyrene-appended imidazolium probe for 2, 4, 6-trinitrophenol in water. *Sensors and Actuators B: Chemical*. 2016 Aug 1; 231: 293-301.

[49] Bains G, Patel AB, Narayanaswami V. Pyrene: a probe to study protein conformation and conformational changes. *Molecules*. 2011 Sep 14; 16 (9):7909-35.

[50] Lewis FD, Zhang Y, Letsinger RL. Bispyrenyl excimer fluorescence: a sensitive oligonucleotide probe. *Journal of the American Chemical Society*. 1997 Jun 11; 119 (23):5451-2.

[51] Betcher-Lange SL, Lehrer SS. Pyrene excimer fluorescence in rabbit skeletal alphaalphatropomyosin labeled with N-(1-pyrene) maleimide. A probe of sulfhydryl proximity and local chain separation. *Journal of Biological Chemistry*. 1978 Jun 10; 253 (11):3757-60.

[52] Ma C, Huang H, Zhao C. An aptamer-based and pyrene-labeled fluorescent biosensor for homogeneous detection of potassium ions. *Analytical Sciences*. 2010 Dec 10; 26 (12):1261-4.

[53] Sahoo D, Weers PM, Ryan RO, Narayanaswami V., Lipid-triggered conformational switch of apolipophorin III helix bundle to an extended helix organization. *Journal of molecular biology*. 2002 Aug 9; 321 (2):201-14.

[54] Goedeweeck M, Van der Auweraer M, De Schryver FC. Molecular dynamics of a peptide chain, studied by intramolecular excimer formation. *Journal of the American Chemical Society*. 1985 Apr; 107 (8):2334-41.

[55] Valanciunaite J, Kempf E, Seki H, Danylchuk DI, Peyriéras N, Niko Y, Klymchenko AS. Polarity mapping of cells and embryos by improved fluorescent solvatochromic pyrene probe. *Analytical chemistry*. 2020 Mar 10; 92 (9):6512-20.

[56] Hazawa M, Amemori S, Nishiyama Y, Iga Y, Iwashima Y, Kobayashi A, Nagatani H, Mizuno M, Takahashi K, Wong RW. A light-switching pyrene probe to detect phase-separated biomolecules. *I Science*. 2021 Aug 20; 24 (8):102865.

[57] Yu C, Yam VW. Glucose sensing via polyanion formation and induced pyrene excimer emission. *Chemical communications*. 2009 (11):1347-9.

[58] Mu YL, Pan L, Lu Q, Xing S, Liu KY, Zhang X. A bifunctional sensitive fluorescence probe based on pyrene for the detection of pH and viscosity in lysosome. *Spectrochimica Acta Part A: Molecular and Biomolecular Spectroscopy*. 2022 Jan 5; 264: 120228.

[59] Chao J, Wang H, Zhang Y, Yin C, Huo F, Sun J, Zhao M. A novel pyrene-based dual multifunctional fluorescent probe for differential sensing of pH and HSO3− and their bio imaging in live cells. *New Journal of Chemistry*. 2018; 42 (5):3322-33.

[60] Chao J, Xu M, Liu Y, Zhang Y, Huo F, Yin C, Wang X. A Pyrene-Based Turn-On Fluorescence Probe for CN− Detection and Its Bio imaging Applications. *Chemistry Select*. 2019 Mar 22; 4 (11):3071-5.

[61] Niu Q, Lan L, Li T, Guo Z, Jiang T, Zhao Z, Feng Z, Xi J. A highly selective turn-on fluorescent and naked-eye colorimetric sensor for cyanide detection in food samples and its application in imaging of living cells. *Sensors and Actuators B: Chemical*. 2018 Dec 10; 276 :13-22.

[62] Li Q, Wang Z, Song W, Ma H, Dong J, Quan YY, Ye X, Huang ZS. A novel D-π-A triphenylamine-based turn-on colorimetric and ratiometric fluorescence probe for cyanide detection. *Dyes and Pigments*. 2019 Feb 1; 161:389-95.

[63] Hossain SM, Prakash V, Mamidi P, Chattopadhyay S, Singh AK. Pyrene-appended bipyridine hydrazone ligand as a turn-on sensor for Cu^{2+} and its bio imaging application. *RSC Advances*. 2020; 10 (7):3646-58.

[64] Choragi A, Mondal J, Manna AK, Chowdhury S, Patra GK. A novel pyrene based highly selective reversible fluorescent-colorimetric sensor for the rapid detection of Cu^{2+} ions: application in bio-imaging. *Analytical Methods*. 2018; 10 (9):1063-73.

[65] Kaur M, Kaur P, Dhuna V, Singh S, Singh K. A ferrocene–pyrene based 'turn-on'chemodosimeter for Cr^{3+} application in bioimaging. *Dalton Transactions*. 2014; 43 (15): 5707-12.

[66] Winnick, FM. *Chem. Rev.*, 1993, 93, 587-614.

[67] Albagli D, Bazan GC, Schrock RR, Wrighton MS. New functional polymers prepared by ring-opening metathesis polymerization: study of the quenching of luminescence of pyrene end groups by ferrocene or phenothiazine centers in the polymers. *The Journal of Physical Chemistry*. 1993 Sep; 97 (39):10211-6.

[68] Saravanan A, Shyamsivappan S, Suresh T, Subashini G, Kadirvelu K, Bhuvanesh N, Nandhakumar R, Mohan PS. An efficient new dual fluorescent pyrene based chemosensor for the detection of bismuth (III) and aluminum (III) ions and its applications in bio-imaging. *Talanta*. 2019 Jun 1; 198 : 249-56.

[69] Jabłoński A, Fritz Y, Wagenknecht HA, Czerwieniec R, Bernaś T, Trzybiński D, Woźniak K, Kowalski K. Pyrene–nucleobase conjugates: synthesis, oligonucleotide binding and confocal bioimaging studies. *Beilstein journal of organic chemistry*. 2017 Nov 28; 13 (1):2521-34.

[70] Ayyavoo K, Velusamy P. Pyrene based materials as fluorescent probes in chemical and biological fields. *New Journal of Chemistry*. 2021; 45 (25):10997-1017.

[71] Wu YS, Li CY, Li YF, Li D, Li Z. Development of a simple pyrene-based ratiometric fluorescent chemosensor for copper ion in living cells. *Sensors and Actuators B: Chemical*. 2016 Jan 1; 222:1226-32.

[72] Wu WN, Mao PD, Wang Y, Mao XJ, Xu ZQ, Xu ZH, Zhao XL, Fan YC, Hou XF. AEE active Schiff base-bearing pyrene unit and further Cu^{2+}-induced self-assembly process. *Sensors and Actuators B: Chemical*. 2018 Apr 1; 258: 393-401.

[73] Rajasekaran D, Venkatachalam K, Periasamy V. "On–off–on" pyrene-based fluorescent chemosensor for the selective recognition of Cu^{2+} and S^{2-} ions and its utilization in live cell imaging. *Applied Organometallic Chemistry*. 2020 Mar; 34(3): e5342.

[74] Chakraborty N, Chakraborty A, Das S. A pyrene based fluorescent turn on chemosensor for detection of Cu^{2+} ions with antioxidant nature. *Journal of Luminescence*. 2018 Jul 1; 199: 302-9.

[75] Phapale D, Gaikwad A, Das D. Selective recognition of Cu (II) and Fe (III) using a pyrene based chemosensor. *Spectrochimica Acta Part A: Molecular and Biomolecular Spectroscopy*. 2017 May 5; 178: 160-5.

[76] Harsha KG, Appalanaidu E, Rao BA, Baggi TR, Rao VJ. ON–OFF Fluorescent Imidazole Derivative for Sensitive and Selective Detection of Copper (II) Ions. *Russian Journal of Organic Chemistry*. 2020 Jan; 56 (1):158-68.

[77] Doganci E. Synthesis, characterization and chemical sensor applications of pyrene side-functional polylactide copolymers. *Polymer International*. 2021 Feb; 70 (2): 202-11.

[78] Harsha KG, Appalanaidu E, Chereddy NR, Baggi TR, Rao VJ. Pyrene tethered imidazole derivative for the qualitative and quantitative detection of mercury present in various matrices. *Sensors and Actuators B: Chemical*. 2018 Mar 1; 256: 528-34.

[79] Walekar LS, Hu P, Molamahmood HV, Long M. FRET based integrated pyrene-AgNPs system for detection of Hg (II) and pyrene dimer: Applications to environmental analysis. *Spectrochimica Acta Part A: Molecular and Biomolecular Spectroscopy*. 2018 Jun 5; 198: 168-76.

[80] Yoon SA, Lee J, Lee MH. A ratiometric fluorescent probe for Zn^{2+} based on pyrene-appended naphthalimide-dipicolylamine. *Sensors and Actuators B: Chemical*. 2018 Apr 1; 258: 50-5.

[81] Thirupathi P, Lee KH. A ratiometric fluorescent detection of Zn (II) in aqueous solutions using pyrene-appended histidine. *Bioorganic & medicinal chemistry letters*. 2013 Dec 15; 23 (24):6811-5.

[82] Zhou Y, Zhu CY, Gao XS, You XY, Yao C. Hg^{2+}-selective ratiometric and "off–on" chemosensor based on the azadiene-pyrene derivative. *Organic Letters*. 2010 Jun 4;12(11):2566-9.

[83] Yang MH, Thirupathi P, Lee KH. Selective and sensitive ratiometric detection of Hg (II) ions using a simple amino acid based sensor. *Organic Letters*. 2011 Oct 7; 13(19): 5028-31.

[84] Liu LN, Tao H, Chen G, Chen Y, Cao QY. An amphiphilic pyrene-based probe for multiple channels sensing of mercury ions. *Journal of Luminescence*. 2018 Nov 1; 203: 189-94.

[85] Bai CB, Xu P, Zhang J, Qiao R, Chen MY, Mei MY, Wei B, Wang C, Zhang L, Chen SS. Long-wavelength fluorescent chemosensors for Hg^{2+} based on pyrene. *ACS omega*. 2019 Aug 29;4(11):14621-5.

[86] Ni XL, Wang S, Zeng X, Tao Z, Yamato T. Pyrene-Linked Triazole-Modified Homooxacalix [3] arene: A Unique C 3 Symmetry Ratiometric Fluorescent Chemosensor for Pb^{2+}. *Organic letters*. 2011 Feb 18; 13(4): 552-5.

[87] Yadav P, Gond S, Singh AK, Singh VP. A pyrene-thiophene based probe for aggregation induced emission enhancement (AIEE) and naked-eye detection of fluoride ions. *Journal of Luminescence*. 2019 Nov 1; 215: 116704.

Chapter 2

Pyrene as a Chromophore in Various Organic Materials

V. Jayathirtha Rao[*]

Fluoro-Agro Chemicals Department and AcSIR-Ghaziabad,
CSIR-Indian Institute of Chemical Technology, Uppal Road Tarnaka,
Hyderabd, Telangana, India

Abstract

Pyrene was identified as one of the aromatic substances from the coal-tar distillation. This aromatic residue, not obeying [4n+2] Huckel`s rule, has been subjected to a variety of research investigations because of its interesting properties. Notably, its light absorption, fluorescence light emission, excimer luminescence, medium polarity sensitive luminescence, quenchable fluorescence, aromatic stacking nature and DNA interactable fluorophore properties are extensively utilized for making innumerable substrates having a variety of optical and optoelectronic applications. This chapter is focused on narrating some basic (i) photophysical properties of pyrene and its derivatives; (ii) pyrene as a core molecule in OLED materials; (iii) pyrene as a molecule in organic solar cell materials; and (iv) pyrene in Non-Linear Optical properties.

Keywords: pyrene, photophysical properties, OLEDs, solar cells, NLO materials, two photon absorption

---

[*] Corresponding Author's Email: vaidya.opv@gmail.com.

In: Pyrene: Chemistry, Properties and Uses
Editor: Charles R. Howe
ISBN: 979-8-88697-670-0
© 2023 Nova Science Publishers, Inc.

Introduction

Pyrene is chosen as a chromophore by several researchers involving a variety of research domains for the past several decades and this is because of its unique properties. Pyrene was discovered in 1837 by Laurent from the destructive distillation of coal tar (Laurent 1837). Pyrene is subjected to extensive chemical transformations leading to the production of several varieties of pyrene derivatives to achieve altercation of desired property to suit the required application (Winnik 1993; Figueira-Duarte and Klaus Mullen 2011). Pyrene is selected as an optical probe in research studies because of its structured UV-Vis absorption, structured fluorescence, formation of excimer and its fluorescence or fluorescence quenching, desirable fluorescence lifetime, acts as nonradiative energy acceptor, its sensitivity to structured UV-Vis absorption and fluorescence towards the environment, and pyrene capacity to associate with surfaces like silica, alumina, zeolites and clays. Pyrene fluorescence is used as a probe to understand the structural aspects of peptides and proteins (Hammarstrom et al. 1997; Sahoo et al. 2000 and 2002; Hajime Maeda et al. 2006). Oligonucleotides, nucleotides and DNA binding studies involving pyrene and its derivatives provide information on the nucleotide sequence, binding sites and nature of binding (Paris et al. 1998; Lewis et al. 1997; Yamana et al. 2002 & 1997; Tong et al. 1995). Lipid structure was probed (Ollman et al. 1987; Pap et al. 1995; Song & Swanson 1999) using lipid anchored pyrene fluorescence, excimer fluorescence, fluorescence quenching and also energy transfer process. The influence of pH, viscosity and temperature (Templer et al. 1998; Birks et al. 1964; Pokhrel et al. 2000) on pyrene fluorescence studies also helps analysis.

There are several advantages to using organic molecules for various optical and optoelectronic applications, like quick and cost-effective preparation, purity of the sample, film forming ability, thermal properties, electrochemical properties, light absorption and luminescence properties further one can tune these properties by attaching suitable substituents and making desirable derivatives. These organic materials are expected to replace inorganic semiconductors to reduce the cost of production, and fabrication of big area devices and further make flexible electronic devices to benefit global consumers. Recent developments indicate that pyrene is being used as a luminophore and also as a semiconductor for various organic optical and optoelectronic applications (Figueira-Duarte and Klaus Mullen 2011). This chapter narrates briefly UV-Vis absorption & Fluorescence properties of pyrene and the main emphasis will be on the utility of pyrene as a chromophore and as a semiconductor in various optical and optoelectronic applications like OLEDs, Solar Cells, NLO and Two-Photon Absorption, Ion Sensors, and as NIR Dyes, covering recent research findings.

Pyrene Photophysical Properties

Pyrene is a flat alternant aromatic compound with four benzene rings (tetracyclic) fused, colourless or pale-yellow colour solid, mp 156°C. Pyrene is a product of the destructive distillation of coal tar. Figure 1 describes pyrene UV-Visible absorption spectra in cyclohexane and fluorescence spectra in ethanol, dichloromethane and acetone solvents. Pyrene absorption spectrum

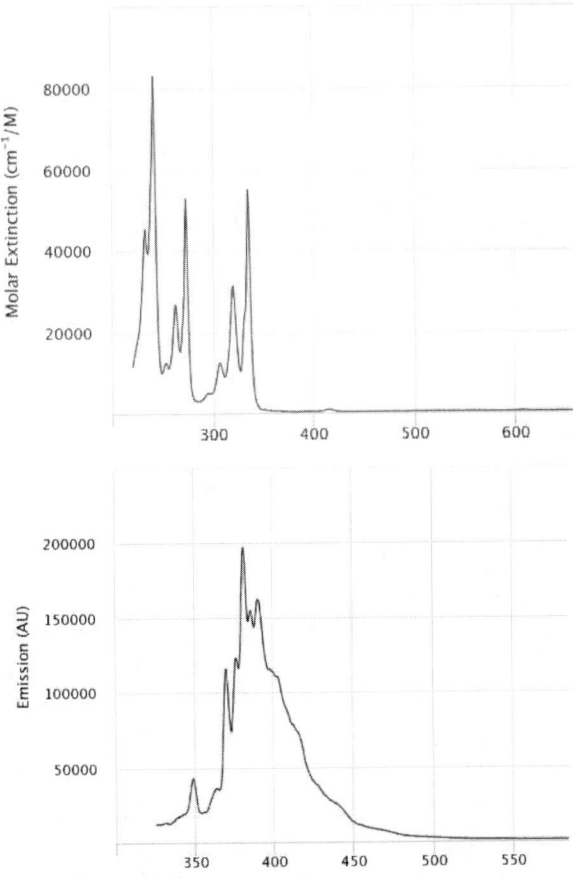

Figure 1. UV-Vis absorption spectrum and fluorescence spectra of pyrene.

indicates that it occupies ~220 nm to ~340 nm spectral range (Clar et al. 1963) with a structured shape. The structured shape informs about the vibrational levels embedded in electronic excitation. The fluorescence spectrum is depicted in Figure 1, covering a spectral region of ~360 nm to ~500 nm. Some structural features appearing in fluorescence spectrum also indicates the involvement of vibrational levels. The fluorescence spectrum is a bit blurred compared to the UV-Vis absorption spectrum (Figure 1).

The first singlet energy level of pyrene is located at ~ 3.5 eV (~370 nm) and the triplet energy level is at 2.1 eV (~590nm) as reported and calculated (Clar and Schmidt 1976). Pyrene fluorescence lifetime (Hautala, Schore & Turro 1973) measured is 370 n sec in cyclohexane (RT) under deaerated conditions and 650 n sec in cyclohexane at rt (Deloius et al. 1979). Phosphorescence lifetime of pyrene (Langelaar, Rettschnick and Hoijtink 1971) is determined for pyrene to be 30 sec at 77°K. Pyrene is known to form excimer and display excimer fluorescence emission (Foster & Kasper 1954), which is broad and structureless emission. The excimer emission of pyrene is pyrene concentration dependent, a simple mechanism is provided in Scheme 1 and is very important for analytical purposes (Ge et al. 2020). Pyrene is an interesting molecule in terms of its triplet-triplet annihilation (Bohne, Abuin, and Scaiano 1990; Scheme 1) phenomenon (TTA) leading to fluorescence. The fluorescence resulting from this TTA is known as delayed fluorescence.

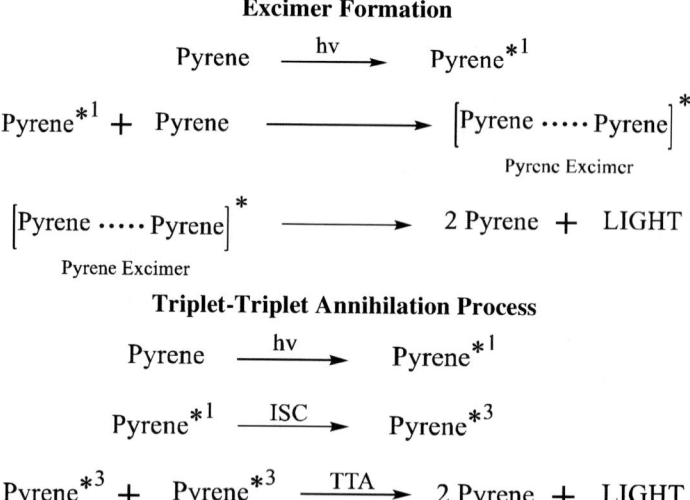

Scheme 1. Describes pyrene excimer formation and triplet-triplet annihilation process.

A highly resolved fluorescence spectrum of pyrene is generated from the pyrene bound over the surface of TLC silica gel and the generated fluorescence line spectrum is similar to the Shoplskii spectrum (Hofstraat et al. 1984). Pyrene monomer and dimer (excimer) emissions were recorded in a zeolitic environment, depending on the loading of pyrene and also available zeolitic space (Liu, Lu & Thomas 1989; Lu & Thomas 1990). The zeolite encapsulated pyrene excited state quenching studies were conducted involving oxygen and Cu+2 ion. Room temperature phosphorescence (Mohan Raj et al. 2019) was recorded for pyrene in an octa acid capsule environment and in the presence of heavy atom Xenon. The electronic and structural properties of pyrene and its derivatives were evaluated by adopting quantum chemical computational techniques (Massimo Ottonelli et al. 2011). Pyrene attached with acrylic moiety displayed fluorescence solvatochromism at rt indicating the formation of intramolecular charge transfer state (but not excimer emission) and the structural features of fluorescence disappeared by changing the solvent hexane to acetonitrile or methanol (Raj Gopal et al. 1997).

Pyrene Moiety

Organic Light Emitting Diodes

Over the past few decades organic light emitting devices (OLEDs) have attracted much attention scientifically and technologically due to their low cost, better resolution, energy efficiency, wide angle viewing and etc. OLEDs

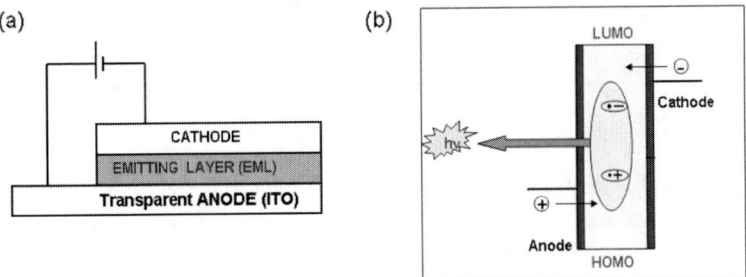

Figure 2. OLED device structure – simple representation.

originating from different modes of luminescence, operational mechanisms, and device structures offer a wide range of challenges in the current research scenario. The conversion of electrical energy into light energy is termed "electroluminescence". Enhancing the electrochemiluminescence of organic molecules has a profound role to play in Organic Light Emitting Devices. Small organic molecules, particularly flat aromatic molecules with luminescence properties are indeed involved with OLEDs. Electrochemiluminescence (ECL) was first reported by Pope et al. in 1963 by passing very large voltages through anthracene crystals (Pope et al. 1963). Later it was a breakthrough informed by Tang & VanSlyke to display a method to fabricate OLEDs (Tang & VanSlyke 1987) using organic substrates. Figure 2 provides a simple representation of OLEDs and it is always interesting to identify the luminescent moiety in these OLEDs and below it is given mechanism of luminescence with many possibilities (Scheme 2).

1) $OM + electron \rightarrow OM^{-\cdot}$ (radical anion)
2) $OM \rightarrow OM^{+\cdot} + electron$ (radical cation)
3) $OM^{-\cdot} + OM^{+\cdot} \rightarrow OM + OM^{*1}$ (singlet excited state)
4) $OM^{*1} \rightarrow OM + h\nu$ (fluorescence)
5) $OM^{-\cdot} + OM^{+\cdot} \rightarrow OM + OM^{*3}$ (triplet excited state)
6) $OM^{*3} \rightarrow OM + h\nu$ (phosphorescence)
7) $OM^{*3} + OM^{*3} \rightarrow OM^{*1} + OM$ (fluorescence: triplet-triplet annihilation)
8) $OM^{*1} + OM \rightarrow [OM....OM]^*$ (singlet excimer)
9) $OM^{*1} + OM2 \rightarrow [OM....OM2]^*$ (singlet exciplex)
10) $OM^{*3} + OM \rightarrow [OM....OM]^*$ (triplet excimer)
11) $OM^{*3} + OM2 \rightarrow [OM....OM2]^*$ (triplet exciplex)
12) $OM^{*1} + OM2 \rightarrow OM + OM2^*$ [Singlet – Singlet Energy Transfer]
13) $OM^{*1} + OM3 \rightarrow OM + OM3^{*3}$ [Singlet - Triplet Energy Transfer]
14) $OM^{*3} + OM4 \rightarrow OM + OM4^{*3}$ [Triplet - Triplet Energy Transfer]
(OM = Organic Molecule: *1 = Singlet; *3 = Triplet)

Scheme 2. Mechanism of luminescence in OLEDs.

Quality functioning of Light Emitting Material become important and some of them are mentioned below as possible criteria to be attended in making suitable organic materials for OLED applications:

1. UV-Vis Absorption
2. Fluorescence - High Luminescence Quantum Yield
3. HOMO-LUMO Energy Gap
4. Melting Point

5. Glass Transition Temperature
6. Amorphous Nature of the Compound
7. Thermal Decomposition Temperature
8. Oxidation and Reduction Potentials
9. Ability to form thin Layers
10. Ability to Conduct Charge
11. Photochemical Stability
12. Radical cation or Radical anion chemistry
13. Colour Purity
14. Generation of Triplet Followed by Efficient Energy Transfer.

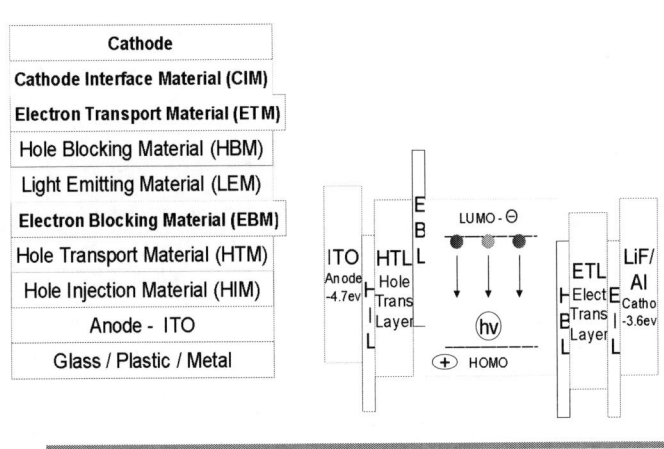

Figure 3. A multilayer OLED structure with several types of organic materials and their functioning property.

Above mentioned properties (criteria) become important for designing new organic materials suitable for OLEDs. The mentioned each property has a role to play in its functioning. It is obvious that satisfying multi property parameters is a real challenge to design and develop new organic materials for OLEDs. Aromatic residues are generally flat molecules, having delocalized π electron clouds, high melting point, good thermal stability, UV-Vis absorption, luminescence, and other properties make them suitable organic materials for OLEDs. This part will high light pyrene based organic materials suitable for OLEDs.

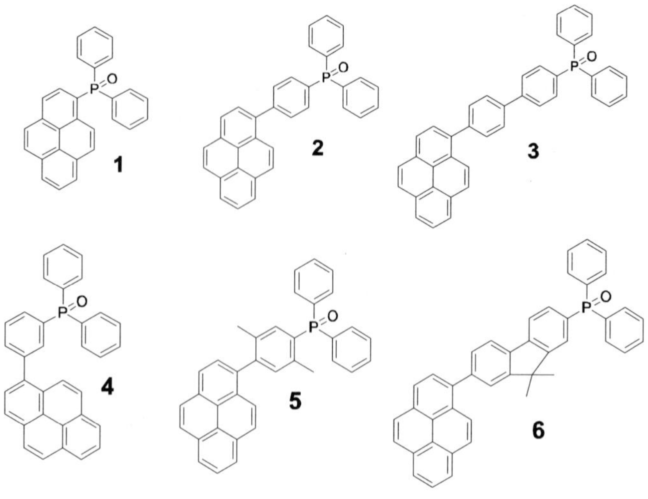

Scheme 3. Pyrene appended phosphine oxides for OLEDs.

Pyrene attached phosphine oxides were synthesized (Mallesham et al. 2015) for OLED fabrication (Scheme 3). These pyrene-appended phosphine oxides are found to be acting as emitting and electron transport materials. Photophysical properties observed for these indicate that there is no change in the UV-Vis absorption in the solution compared to the thin film state, whereas there is a red shift in the luminescence of these molecules comparing solution to film state and the redshift observed in the film state for these **1-6** compounds indicating intermolecular interactions in the thin film state. Indeed compounds **1, 4** and **5** displayed concentration-dependent new emission in the solution phase attributable to "excimer" emission (Figure 4). The solid-state crystal structure determined for compound **2** informs the proximity of two pyrene moieties within the excimer formation distance. Thermal and electrochemical properties evaluated for **1** to **6** are found to be very good for OLED fabrication. The OLED fabrications indicated that these pyrene derivatives **1** to **6** are excellent emitters (Table 1) and also electron transporting materials. The OLED emission properties evaluated (Table 1), inform that intensity, current efficiency, power efficiency and EQE numbers are excellent. The light emitting species of OLEDs fabricated are identified as pyrene residue of compounds **1** to **6**. The phosphine oxide part in these **1** to **6** is acting as an electron transport and energy transfer agent. The high-efficiency numbers explained that phosphine oxide triplets (Soon Ok Jeon and Jun Yeob Lee

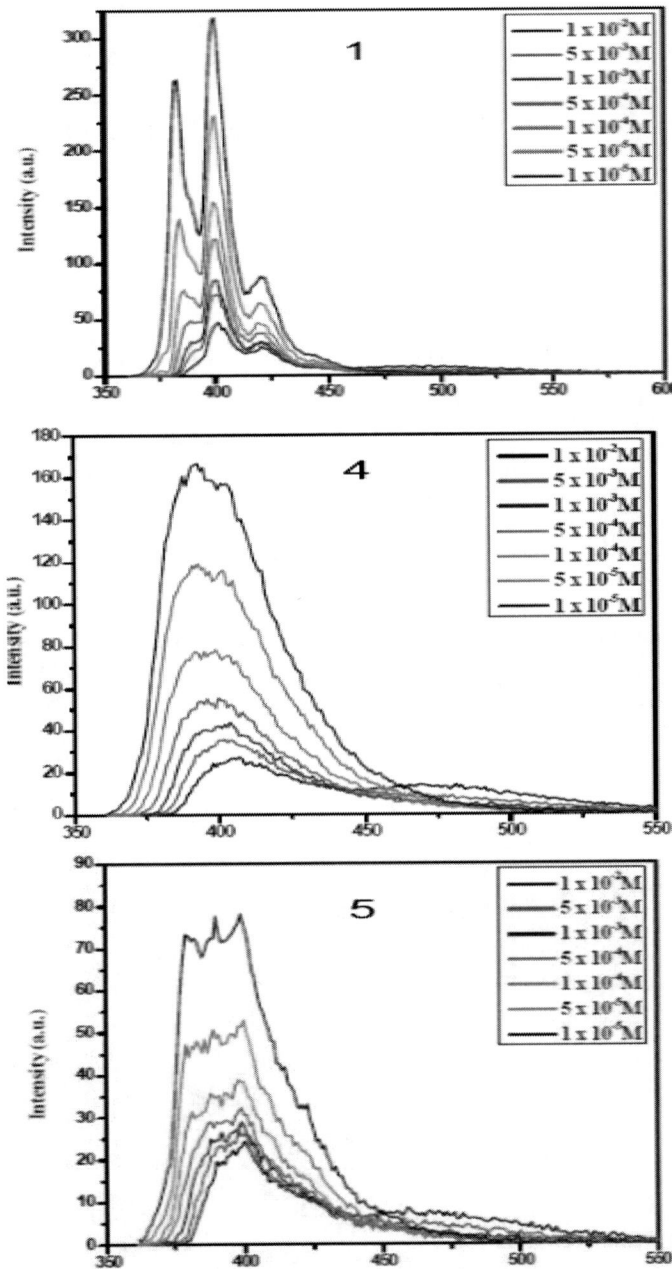

Figure 4. Excimer emission by varying the concentration of pyrene compound **1**, **4** and **5**.

2012) generated, by applying a voltage to fabricated OLEDs, are involved in transferring energy to the pyrene part, there by increasing the luminescence yields. The delayed fluorescence observed at room temperature and at low concentrations of **1** to **6**, is in favour of the triplet to singlet energy transfer process. The mechanism of triplet to singlet energy transfer (harvesting triplets to singlet) indicated from these investigations (Mallesham et al. 2015) has greater implications on the design and OLED efficiency.

Table 1. OLED Emission properties for 1 to 6

Comp.	Intensity Cd/m^2	Current Efficiency Cd/A	Power Efficiency Lm/W	EQE	λ ECL
1	28,500	22.8	11.94	7.4	491 nm
2	34,300	21.1	11.0	7.2	480 nm
3	42,750	24.45	12.8	7.9	476 nm
4	37,050	24.7	12.94	8.0	467 nm
5	34,200	30.1	15.76	9.1	465 nm
6	34,500	24.9	13.0	8.1	473 nm

Pyrene attached with oxadiazole moiety compounds were synthesized (Scheme 4) to understand the mechanism of electrochemiluminescence observed in OLEDs (Swetha et al. 2016).

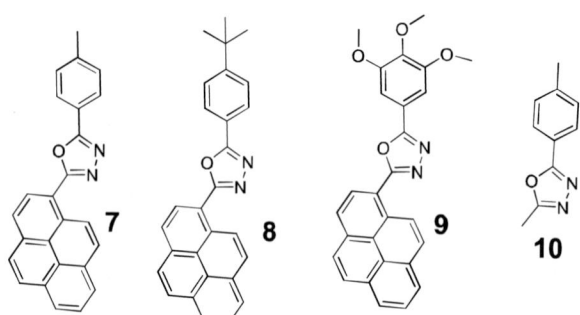

Scheme 4. Pyrene oxadiazoles **7, 8, 9** and 4-MePhenyl-5-methyloxadiazole for OLEDs.

Compounds **7, 8, 9** and fragment **10** were synthesized, and then their material properties were determined. UV-Visible absorption spectra of these **7, 8** and **9** are redshifted relatively to the pyrene absorption spectrum. Compounds **7, 8** and **9** absorption spectra are further ~10 nm red-shifted in their film state. Pyrene oxadiazoles **7, 8** and **9** fluorescence spectra show 420 nm to 460 nm and their film state fluorescence is further red-shifted to 510 nm to 530 nm. The fluorescence spectral shape and the redshift observed in the thin film states of **7, 8** and **9** is an indication of intermolecular interactions in the film state. Fluorescence lifetimes measured for **7, 8** and **9** indicate, that these molecules have a shorter lifetime compared to pyrene molecules (Lambart et al. 2012; Crawford et al. 2011) as mentioned earlier (Pati et al. 2015). These pyrene oxadiazoles **7, 8** and **9** exhibited delayed fluorescence on a microsecond time scale in dilute solutions (10^{-5} M) and also in thin films. The origin of delayed fluorescence was attributed to intramolecular oxadiazole triplet energy transfer to singlet pyrene. UV-Vis absorption, fluorescence, phosphorescence and phosphorescence lifetime measurements were determined for fragment compound **10**. The phosphorescence spectral overlap of **10** with the pyrene fluorescence in compounds **7, 8** and **9** (spectral overlap is given in Figure 5) is an indication of possible triplet (compound **10**) to singlet energy transfer (compounds **7, 8** and **9**). The observable delayed fluorescence in compounds **7, 8** and **9** at low concentrations (10^{-5} M) rules out the triplet-triplet annihilation mechanism. Further, it was argued that the charge transfer nature of the excited state and also closely lying excited states (Raj Gopal et al. 1995) do play in this triplet to singlet energy transfer. The thermal properties of these compounds **7, 8** and **9** are found to be excellent and suitable for OLED fabrications.

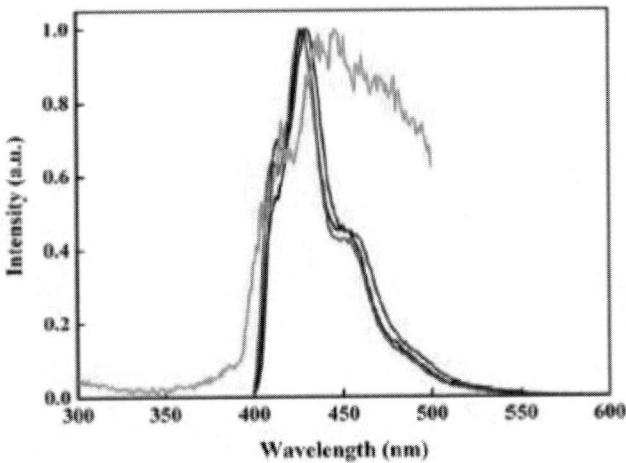

Figure 5. Overlap spectra of phosphorescence of oxadiazole fragment 10 (300 to 500 nm) with fluorescence of 7, 8, and 9 (400 to 550 nm).

Table 2. ECL performance of OLEDs using 7, 8 & 9. Device configuration – ITO(120nm)/F4TCNQ(40nm)/7 or 8 or 9/BCP(6nm)/LiF(1nm)/Al(150nm)

Comp.	V_{onset}	I_{max} cd/m^2	Current Effi. cd/A	Power Effi. Lm/W	External QE	λ ECL
7	2.93	6387	13.06	8.34	5.14	522 nm
8	2.76	8524	15.13	10.95	5.96	518 nm
9	2.61	7050	14.13	9.46	5.49	527 nm

Electrochemiluminescence data for 7, 8 and 9 are arranged in Table 2. The intensity, current efficiency, power efficiency and external quantum efficiency numbers are very good under simple device fabrication conditions. The observed efficiencies for these 7, 8 and 9 are attributed to harvesting triplets formed electrochemically to singlets via the triplet to singlet energy transfer process. In other devices fabricated informed that these 7, 8 and 9 are acting as electron transport (comparing with standard Alq$_3$) and green emitting materials.

Pyrene-Imidazoles 11 to 16, were synthesized (Umashankar et al. 2021) along with three substituted imidazole fragments 17, 18 and 19 (Scheme 5). Pyrene moiety is an emitter and the imidazole part acts as electron transporting and energy transfer species in these 11 to 16 compounds. UV-Vis absorption of 11 to 16 in film state is ~6 nm red-shifted compared to solution-phase absorption maxima. Emission maxima (~432 nm) of 11 to 16 in solution are red-shifted by ~20 nm to 34 nm in film state. Delayed fluorescence was observed for all of the 11 to 16 compounds. The origin of this delayed fluorescence may be intramolecular imidazole triplet energy transfer to singlet pyrene moiety. The measured delayed fluorescence lifetimes of 11 to 16 fall in the range of 7.1 to 8.6 micro sec. Imidazole fragments 17, 18 and 19 prepared were evaluated for their optical properties, particularly phosphorescence spectra generated to determine triplet energies of these and found that they lie in the range of 2.85 to 2.9 eV. Then overlap of phosphorescence spectrum, with the film state fluorescence spectrum of the matched compound with fragment indicate the suitability of triplet energy of imidazole, transfer to singlet pyrene (Figures 6a, 6b and 6c). The high triplet energies of these imidazole-related compounds are found to be comparable theoretically (Daniel Sylvinson et al. 2019) and experimentally (Muazzam Idris et al. 2019).

OLED device fabrications were conducted on these **11** to **16** compounds adopting device configuration as ITO (120 nm)/F$_4$TCNQ (6 nm)/TAPC (30 nm)/Emissive Material (35 nm)/TPBI (30 nm)/LiF (1 nm)/Al (150 nm) and the data is provided in Table 3. All the compounds **11** to **16** exhibited ECL (434 nm to 465 nm) in the blue region. Luminescence, current efficiency, power efficiency and EQE numbers generated are very good compared with other published pyrene-imidazole systems (Liu 2016; Chen 2021). The higher efficiencies observed, reflects the idea of intramolecular triplet energy transfer from imidazole to the pyrene singlet state.

3,3`-di-t-butyl-Carbazole attached at 2 & 7 positions of pyrene - **20**, was synthesized (Chercka et al. 2014; Scheme 5) and its material properties were evaluated. OLED fabrication provided that it is an efficient deep blue ECE emitter at 412 nm with high efficiency. The triplet energy of carbazole ~3.05 eV (Klemens Brunner 2004) plays an important role in transferring triplet energy to form singlet pyrene and followed by emission.

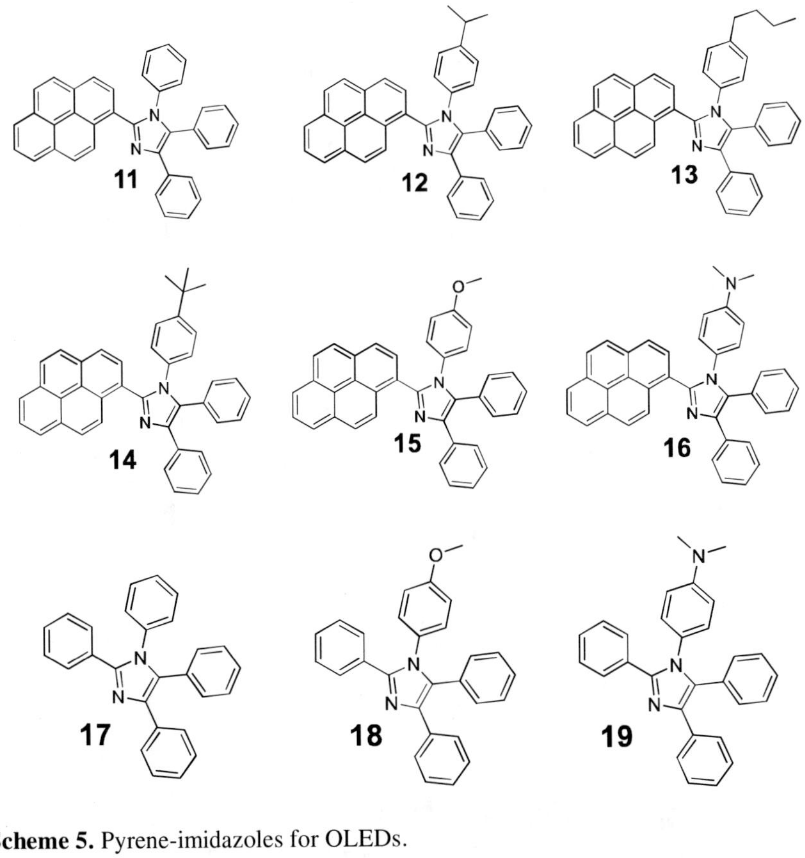

Scheme 5. Pyrene-imidazoles for OLEDs.

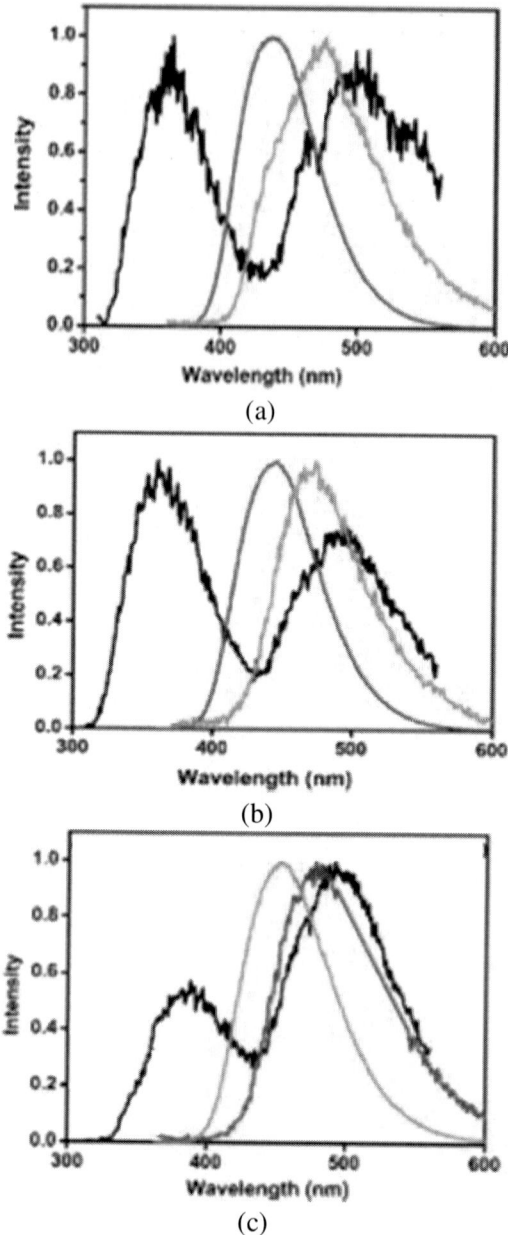

Figure 6. (a) Overlap spectra of **17** phosphorescence with **11** fluorescence in solution and fluorescence in thin film. (b) Overlap spectra of **18** phosphorescence with **15** fluorescence in solution and fluorescence in thin film (c) Overlap spectra of **19** phosphorescence with **16** fluorescence in solution and fluorescence in thin film.

Table 3. Electrochemiluminescence data generated for 11 to 16

Compd.	CBP wt%	V_{onset} V	L_{max} cd m^{-2}	Current Efficiency cdA^{-1}	Power Efficiency lmW^{-1}	EQE	λ ECL
11	nil	4.81	3715	4.73	3.37	1.87	465
11	5%	5.26	10355	9.24	8.07	4.39	447
12	5%	5.19	11428	9.82	8.32	4.64	436
13	5%	5.17	9884	8.66	7.34	3.97	434
14	5%	5.22	9928	8.39	7.23	4.09	438
15	5%	5.14	9174	7.47	6.81	3.48	443
16	5%	5.35	9328	7.81	6.95	3.61	459

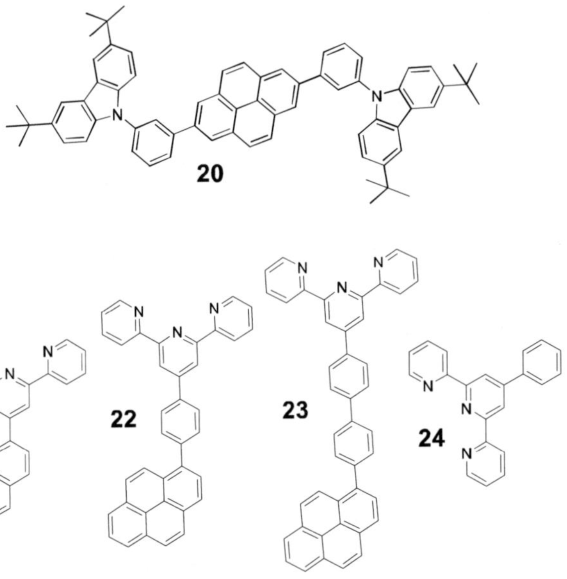

Scheme 6. Pyrene-carbazole and pyrene-terpyridyls for OLEDs.

Scheme 6 illustrates terpyridine attached pyrene – **21, 22** and **23** and phenyl-terpyridine **24** molecules were synthesized (Umashankar et all 2020) and their material properties determined. All the compounds – **21, 22** and **23** exhibited delayed fluorescence at room temperature indicating the possible intramolecular triplet energy transfer from terpyridyl fragment to form singlet pyrene. The phosphorescence spectrum of terpyridine recorded was overlayed with fluorescence spectra of – **21, 22** and **23** to observe a neat overlap and this is in support of intramolecular energy transfer. Pyrene attached to pyridine compounds were prepared (Krishan Kumar et al. 2021), their material properties determined and OLEDs were fabricated to understand them as good hole transporting materials.

Pyrene-Imidazole-Fluorene hybrid compounds were synthesized as a blue emitting (Guowei Chen et al. 2021) molecules for OLED fabrication. Compounds **25** and **26** displayed 448 nm and 444 nm as ECL upon OLED fabrication. EQE numbers are also found to be good. These compounds also inform that there is a possibility of intra-molecular fluorene triplet transfers energy to singlet pyrene-imidazole to observe fluorescence. Further, these compounds can demonstrate delayed fluorescence also. The observed quantum yield of fluorescence is the summations of prompt and delayed fluorescence.

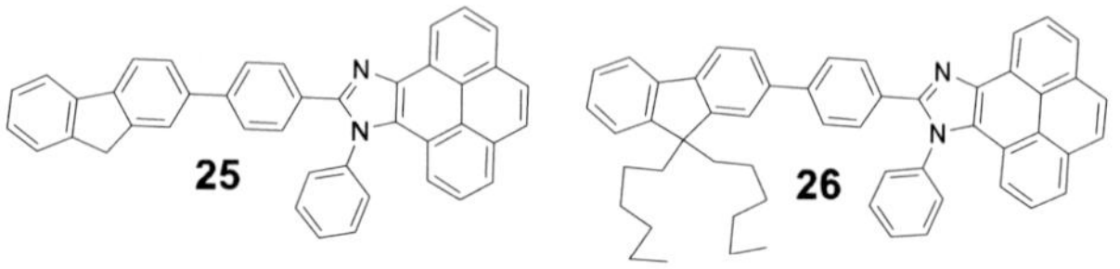

Scheme 7. Pyrene-imidazole-fluorene hybrids for OLEDs.

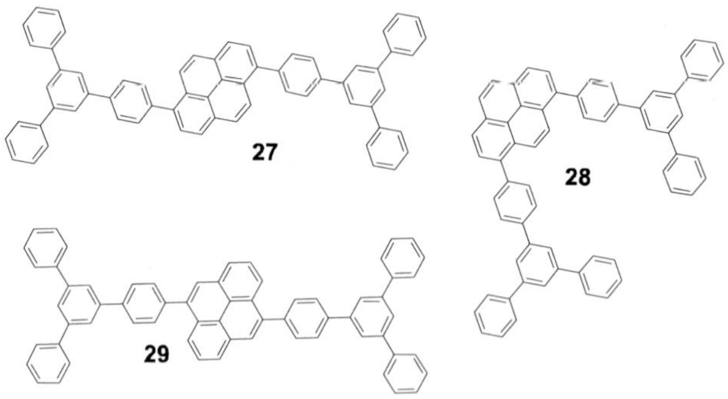

Scheme 8. Suitably substituted pyrenes for OLEDs.

Suitably substituted pyrenes were synthesized (Jung et al. 2018) for blue-emitting (Scheme 8) OLEDs. Symmetrical triphenyl-benzene was linked to pyrene at selected positions (**27, 28** and **29**), as shown in Scheme 7. All the three compounds **27, 28** and **29**, exhibited blue ECL 455 nm, 433 nm and 456 nm respectively. Among the compounds (**27, 28** and **29**) synthesized, one compound **29** showed a good EQE of 7.1 indicating the importance of the site of attachment to the pyrene.

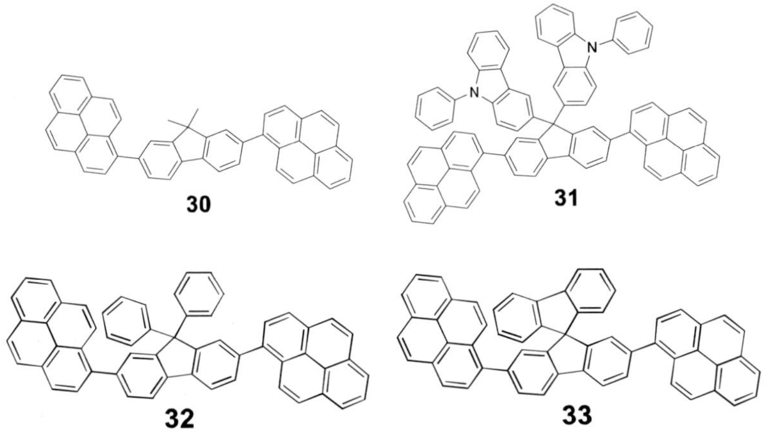

Scheme 9. Pyrene-fluorene, pyrene-spirofluorene and pyrene-fluorene-carbazole hybrids for OLEDs.

Scheme 9 provides pyrene-fluorene – **30, 32**, pyrene-spiro-fluorene **33** and pyrene-fluorene-carbazole – **31,** hybrids for blue emitting OLEDs (Silu Tao et al. 2005 and 2010). The four compounds were synthesized, material characterization data generated and OLEDs fabricated. All the compounds **30, 31, 32** and **33** recorded ECL at 468 nm, 458 nm, 468 nm and 472 nm respectively and also showed good efficiencies. The authors could not report the delayed fluorescence and did not discuss about the possible fluorene or carbazole triplet energy transfer to pyrene singlet.

Adachi and co-workers (Lin-Song Cui et al. 2017; Kenichi Goushi and Chihaya Adachi 2012) introduced TADF – Thermally Activated Delayed Fluorescence idea in OLEDs to improve its efficiency. Close lying triplet excited state transfers energy to the singlet excited state leading to improved fluorescence and this process is activated by thermal means (TADF). We believe that thermal activation is not required when the triplet state is in a higher energy level than the transferable singlet state and also the fabricated OLED devices are held at temperatures ~60 to 100°C, because of resistance created by the organic thin layers sandwiched between two electrodes upon applying voltage. The energy transfer between the triplet excited state to the singlet excited state might have vibronic coupling. This vibronic coupling is further improved by the charge transfer nature of the excited state.

Organic Solar Cells

Solar Energy coming from SUN is available all the time for the global earth residents, it is a renewable form of energy and can be utilized for all purposes. Solar energy is green, clean and renewable reaching the earth in abundance. Earth is exposed to 174 Peta W of SUN energy in form of UV light (8%), Visible light (46%) and infrared light (46%). These facts made researchers find out means of the conversion of SUN light energy into useful forms of electrical energy, where the researchers are heavily engaged in this process of conversion SUN light to electricity. This has become a potential area of research to devise solar energy trapping units into a usable form of electricity. A solar Cell is a device or unit which converts light into electricity. There is a tremendous effort by global researchers to achieve higher and higher solar energy to electricity conversion efficiency by fabricating Solar Cell units and testing them regularly for their photo-conversion efficiency. There is a limitation exists that silicon solar cells reached 26% efficiency and there is difficulty in further improvement. This provided a window to work rigorously on the organic solar cells. A brief note dealing with terminology followed by thin film organic solar cells involving pyrene moiety will be discussed below.

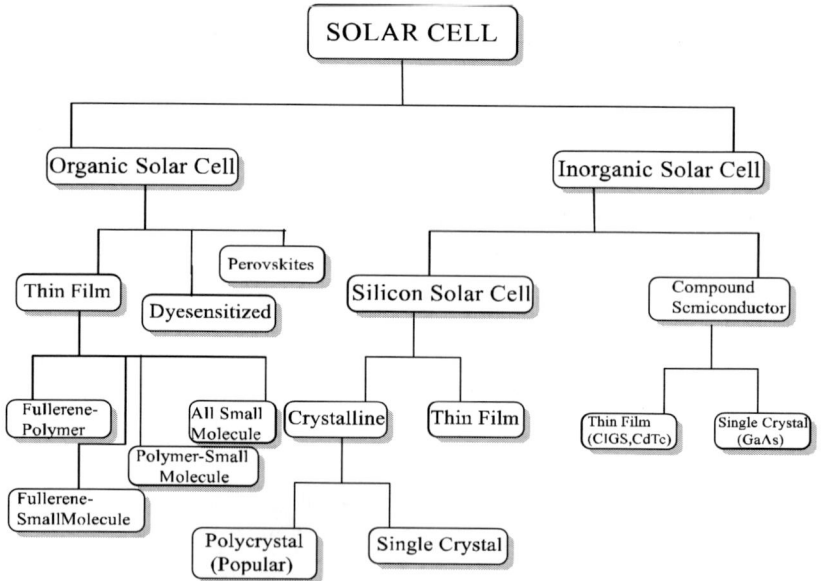

Figure 7. Various types of solar cells.

Various types of solar cells are depicted in Figure 7. Inorganic solar cells are not taken for discussion. Thin film organic solar cells are discussed in this part. There are four types of thin film organic solar cells (Figure 7). The general configuration or the structure of the organic solar cell is given in Figure 8. An active layer consists of donor and acceptor compounds capable of absorbing visible light and sandwiched between anode and cathode. Absorption of light by the active layer produces charges and mobility of these charges towards cathode and anode produces electricity. Figure 9 depicts a simple route of functioning of bulk heterojunction organic solar cells, where light absorption is the initial event and charge extraction is the final event. Some of the technical terms used are defined as follows. (1) Open Circuit Voltage - Voc: The voltage at which no current flows through a solar cell is called open circuit voltage (Voc) and it is the maximum voltage available from the solar cells. (2) Short Circuit Current - Jsc: For V = 0, only the short circuit current (Jsc) flows through the solar cell. Jsc represents the maximum current that could be obtained in a solar cell. This Jsc depends on the number of absorbed photons, the surface area of the photoactive layer, device thickness and charge transport properties of active material. (3) IPCE – Incident Photon-to-Current Efficiency: The incident photon-to-current efficiency is defined as the ratio of the number of incident photons and the number of photo-induced charge carriers which can be extracted out of the solar cell. (4) Power Conversion Efficiency – PCE: It is a measure of the quality of the cell which provides evidence of how much power the cell will generate per incident photon. (5) Fill Factor – FF: The FF, which determines the quality of solar cell can be obtained from the ratio of the maximum power output to the product of its Voc and Jsc and is always <1.

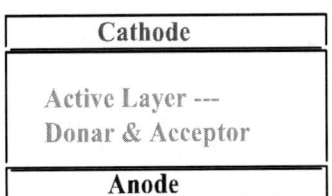

Figure 8. Simple structure of organic solar cell or bulk heterojunction organic solar cell.

Some of the thumb rules can be adopted in making materials and fabricating organic solar cells are listed below.

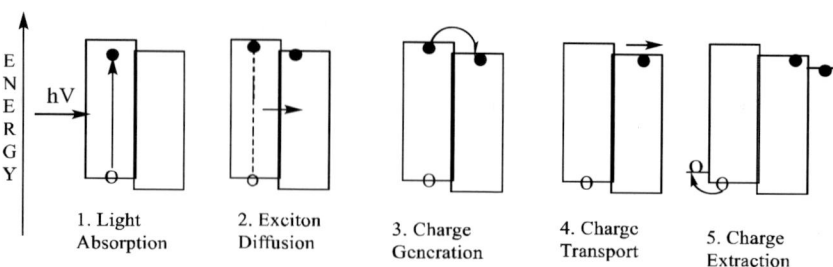

1. Light Absorption
2. Exciton Diffusion
3. Charge Generation
4. Charge Transport
5. Charge Extraction

Figure 9. Simple route of functioning of organic solar cell or bulk heterojunction organic solar cell.

The UV-Visible light absorption property of the donor and acceptor molecules (blend materials) to be used in organic solar cells, should overlap with the solar spectrum and also have high molar absorption coefficients and excellent width at half height of the absorption spectrum. Further, these donor and acceptor materials can maintain complementarity in their light absorption property.

The HOMO and LUMO levels of the donor-acceptor molecules must be properly tuned to observe the photochemical electron transfer upon light absorption.

To obtain a high open-circuit voltage (V_{OC}), a deep HOMO level is to be adjusted, since the maximum value of the V_{OC} is determined by the energy difference between the HOMO level of the donor and the LUMO level of the acceptor.

Charge transport is controlled by balanced charge (negative and positive) mobility of the blend employed in the organic solar cell fabrications.

The solubility parameter of the solar cell organic materials giving to blend, its film forming ability and admixing of donor-acceptor materials providing an acceptable blend morphological structure.

Control of the blend`s supramolecular structure can be introduced by understanding the role of recognition points, leading to good film morphology, tuning UV-Vis absorption and further influencing the charge transport or charge mobility.

Thermal stability (T_d), phase transition temperature (T_g) and crystallinity data of the organic molecules (blend materials) to be used in organic solar cells, needs attention.

This part will be dealing with pyrene attached/appended/carrying organic blend materials and their utility in organic solar cells.

1,4-Diketo-3,6-dithienylpyrrolo[3,4-c]pyrrole (DPP) and Pyrene conjugates are linked with acetylene and were synthesized (Jeong-Wook Mun et al. 2013) to use them in fabricating organic solar cells and also to determine efficiency. The role of the acetylene link between pyrene and DPP is to bring changes in HOMO – LUMO levels and also to make the compound a better donor. Three compounds were synthesized (Scheme 10), two compounds – **34** and **35** with acetylene link and one compound - **36** without acetylene link. Device configuration was ITO/PEDOT-PSS/**34** or **35** or **36** + PC70BM blend. Photovoltaic data generated is in Table 4 informs that when pyrene is linked to DPP via acetylene linkage has a prominent effect on the efficiency of the solar cell functioning.

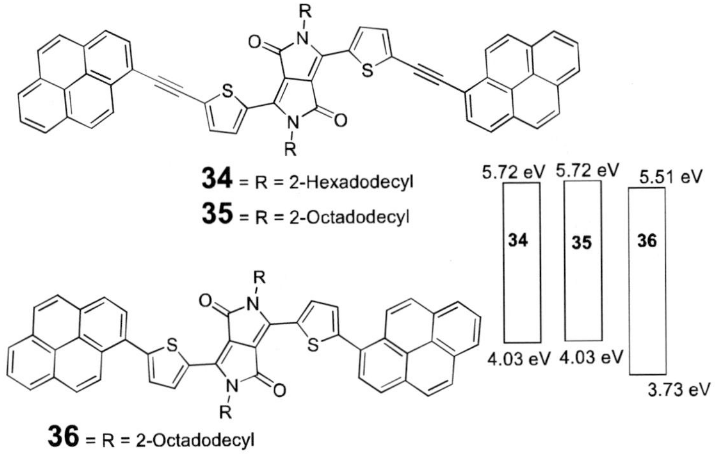
R
O
N
S
S
N
O
R
34 = R = 2-Hexadodecyl
35 = R = 2-Octadodecyl
5.72 eV
5.72 eV
5.51 eV
34
35
36
4.03 eV
4.03 eV
3.73 eV
R
O
N
S
S
N
O
R
36 = R = 2-Octadodecyl

Scheme 10. DPP – Pyrene linked with acetylene compounds.

Table 4. Photovoltaic properties of 34, 35 and 36

Compd.	Jsc	Voc	FF	PCE
34	6.2	0.87	41.9	2.25
35	8.89	0.85	41.7	3.15
36	2.38	0.79	27.2	0.51

Pyrene linked with polymer was synthesized (Ji-Hoon Kim et al. 2012) to understand the role of pyrene on the functioning of organic solar cells. Synthesized polymers **37** and **38** were used for fabricating organic solar cells with the given cell configuration: ITO/PEDOT-PSS/**37** or **38** + PC71BM/ Ca/Al. The photovoltaic information gathered is arranged in Table 6. Indeed the efficiency of an organic solar cell fabricated using **37** and **38** is very good and also higher when compared to non-pyrene attached polymers. The rationale given by authors for the observations made is that attaching pyrene to polymer improves thermal properties, and improves hole mobility due to π − π interactions, as the pyrene content increases the hole mobility process also improves and also as the pyrene content increases the HOMO levels found decreasing.

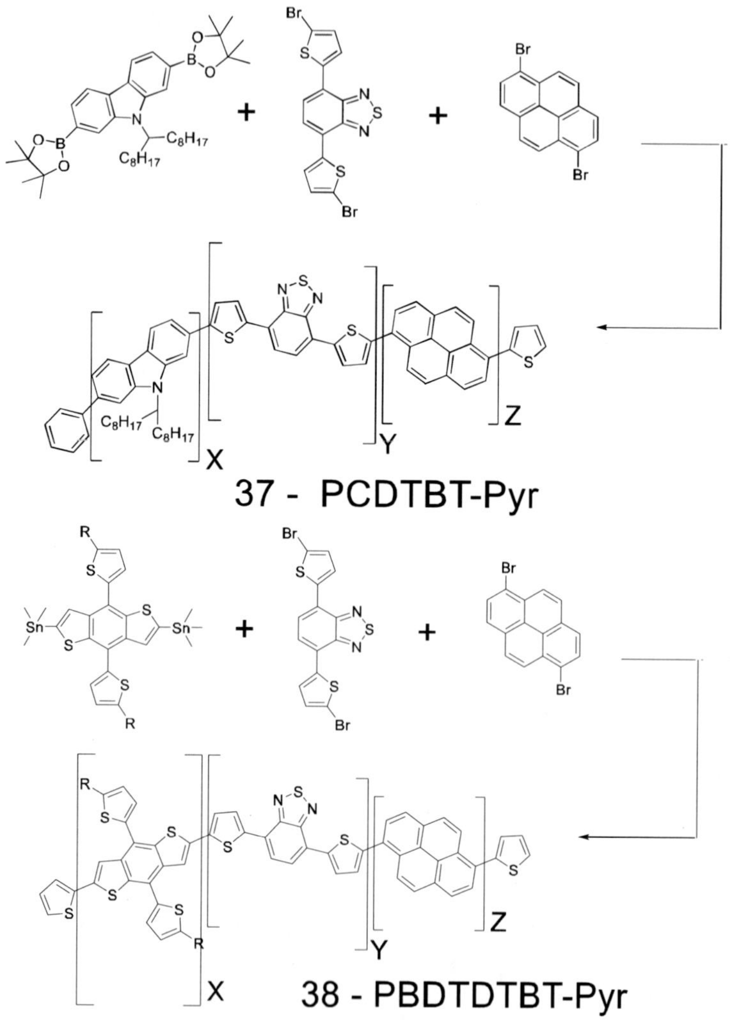

Scheme 11. Pyrene linked with polymer for organic solar cells.

Table 5. Photovoltaic properties of 37 and 38

Compound	Jsc	Voc	FF	PCE
37 PCDTBT-5Pyr	10.60	0.74	41	3.22
38 PBDTDTBT-10Pyr	10.93	0.87	53	5.04

A pyrene-bodipy compound **39** was prepared (Anastasia Soultati et al. 2020) and used as a layer for a non-fullerene organic solar cell fabrication to understand the importance of interfacing properties. Organic solar cell fabricated with configuration: Glass/FTO/ZnO or ZnO/Py-BDP/PM6:IT-4F (**40**)/MoOx/Al. The device characteristics developed are placed in Table 6. Placing the pyrene-bodipy compound **39** as an interfacing layer in organic solar cell fabrication affected the reduction of Wf (workforce) of ZnO and improved the electron transport property. Pyrene-bodipy **39** has improved film morphology, by involving π- π, dipole-dipole and π-dipole and other supramolecular interactions in a thin film, thereby bringing "stacking" of molecules of the organic solar cell fabricated. These improved morphological features paved the way for the higher efficiency numbers (Table 6) like 11.80%. The control experiment conducted without the pyrene-bodipy **39** gave only 9.86% efficiency (Table 6) indicating the importance of the interfacing layer. More interestingly, the pyrene-bodipy **39** interface layer is found to protect the donor-acceptor blend photostability or inhibit the photodegradation of blend molecules, leading to better stability of the organic solar cells.

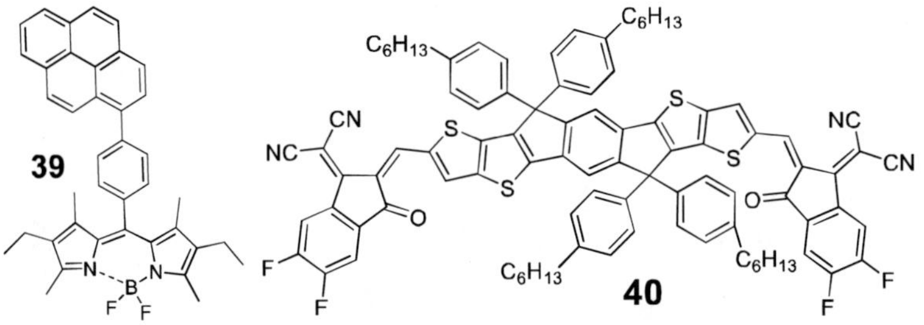

Scheme 12. Pyrene linked with bodipy and compound **40** for organic solar cells.

Table 6. Photovoltaic properties of 40 in the presence of 39

Compound	Jsc	Voc	FF	PCE
40 without 39	19.14	0.80	68	10.41
40 with 39	19.98	0.82	72	11.80

Pyrene core-based, contiguously fused aromatic and heteroaromatic ring compounds **41** and **42** (Shungang Liu et al. 2020) were synthesized for the purpose of fabricating and to understand the efficiency of organic solar cells. Scheme 13 provides structural aspects with HOMO – LUMO values. ITO/ZnO/PFN-Br/PTB7-Th:**41** or **42**/MoO3/Al is the configuration of the organic solar cell fabricated. PTB7-Th is a polymer employed as a donor in the blend with compounds **41** or **42**. Authors employed di-phenylethane, 1,8-diiodooctane and 1-chloro-naphthalene used as additives to fine-tune the morphology of the film and indeed 1-chloro naphthalene improved the film morphology as indicated in the increased efficiency. The photovoltaic performance data is given in Table 8. The compound **42** did not give the comparable efficiency upon fabrication of organic solar cell and was termed as 40% inefficient compared to **41**. The main difference between **41** and **42** is the connectivity of rings with the pyrene core position and also the six or five-membered ring present in **41** and **42**. The suitable connectivity of rings makes (contiguously fused ring structure) a difference in their UV-Vis absorption, which depends upon their positive overlap of orbitals (Chetti Prabhakar et al. 2005).

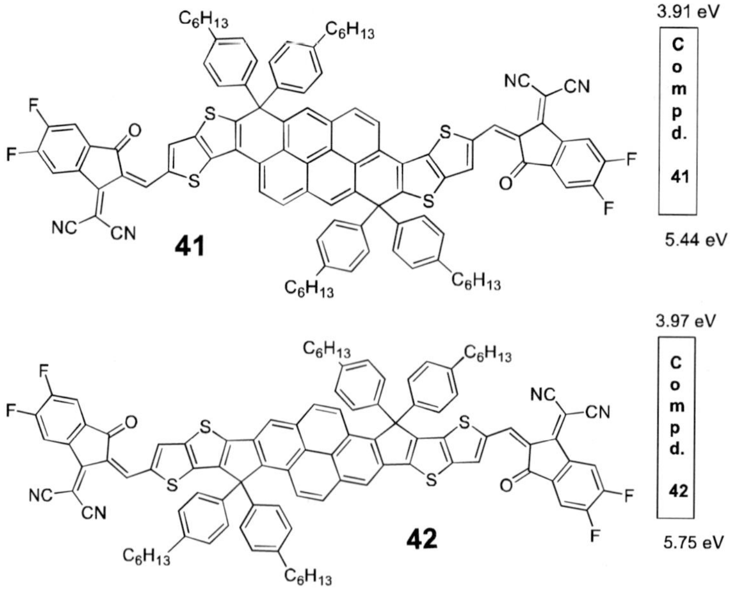

Scheme 13. Pyrene-core based contiguously fused ring compounds for organic solar cells.

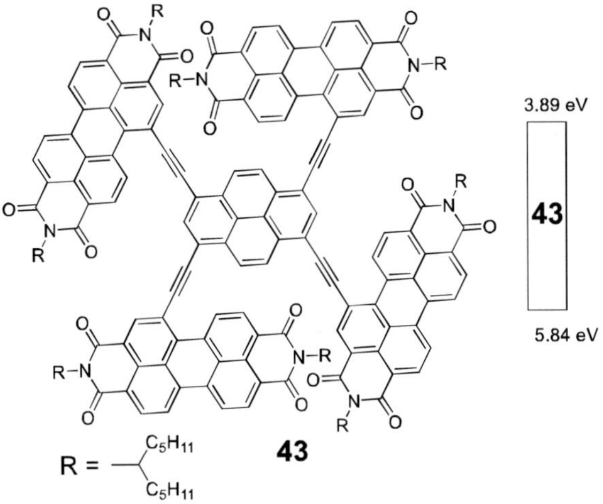

Scheme 13a. Pyrene – perylene diimide linked with acetylene compound **43** for organic solar cells.

Table 7. Photovoltaic properties of 40 in the presence of 39

Compound	Jsc	Voc	FF	PCE
41 with 1-CN	21.5	0.79	67	11.55
41 without 1-CN	20.16	0.79	58	09.11

Table 8. Photovoltaic properties of 43

Blend	Jsc	Voc	FF	PCE
43 (1.5) + 1-CN (3%) + PTB7-Th (1.0)	18.13	0.78	54.48	07.71
PTB7-Th (1.0) + ITIC (1.13)	15.60	0.81	60.89	07.73
43 (0.07) + PTB7-Th (1.0) + ITIC (1.17)	18.18	0.79	60.32	08.76
43 (0.05) + PTB7-Th (1.0) + ITIC (0.95) + DIO (0.5%)	18.50	0.88	66.64	10.93

Further, the authors speak that a 1,8-positional link with pyrene is a crucial point in improving the efficiency of organic solar cell performance.

Four perylene-diamide molecules attached to pyrene core with acetylene linker **43** were synthesized (Feng Tang et al. 2019) for the purpose of fabricating organic solar cells to understand the linker acetylene's role. Organic solar cells were fabricated using binary and ternary blends. The photovoltaic parameters obtained are in Table 8. Compound **43** as an acceptor and PTB7-Th polymer as donor blend gave 7.71% efficiency. Another blend where ITIC (0.95) as acceptor, PTB7-Th (1.0) as donor, DIO (0.5%) and **43**

(0.05) as enhancer provided 10.93% efficiency. Authors concluded that synthesized compound 43 acts as an acceptor and also when it is mixed in a blend with lesser composition, it enhances efficiency. The rationale provided that the "acetylenic" link is helping to mobilize charge mobility/migration. This argument is supported by the charge transport properties determined for the ternary blend used for organic solar cell fabrication and the number determined is 1.2 x 10^{-4} (μ_e (cm^2 V^{-1} s^{-1}). Present studies indicate that the pyrene-acetylene combination having good π orbital cloud spacers will improve the efficiency in the case of non-fullerene acceptors.

Organic Nonlinear Optics

Research deals with intense light interaction with organic materials can be known as Organic Nonlinear Optics. The intense light of a laser beam strikes an atom or a molecule, it manifests in electrical polarization in the molecule or atom, because of movements of electrons around the influence of the nucleus. This high-intensity laser light-induced polarization in molecules or atoms presents properties optically non-linear. Thus, the created polarization at the molecular level has the capacity to modify to second, third and even to higher harmonics. The magnitude or degree of polarization in the molecule, resulting from intense laser light, will determine the nonlinear optical property of the molecule. The incredible growth in the area of internet, multimedia and information technology has its roots in NLO materials. Data control cloud computing, AI (artificial intelligence and the number of people using mobile phones (Globally) are rapid upcoming domains. The present Global scenario provides incredible growth (Technical Development 2018) and innumerable opportunities for the use of optoelectric and electro-optic devices to process, transmit, and store information faster and better (Marpaung et al. 2019; Jiang et al. 2018) and where the phenomenon of Nonlinear Optics plays a pivot role. Organic NLO materials have several advantages, quite a few applications and are now popular. Organic materials are more important for NLO studies because, they are cheaper, and simple to fabricate, easy to fine-tune properties at the molecular level, stability towards laser beam, smaller dielectric constants, good thermal & mechanical properties, light absorption property of the organic molecule, film forming ability, large area of π conjugated system and relatively better laser response time. The following narration is limited to the organic nonlinear materials involving pyrene residue of recent citations.

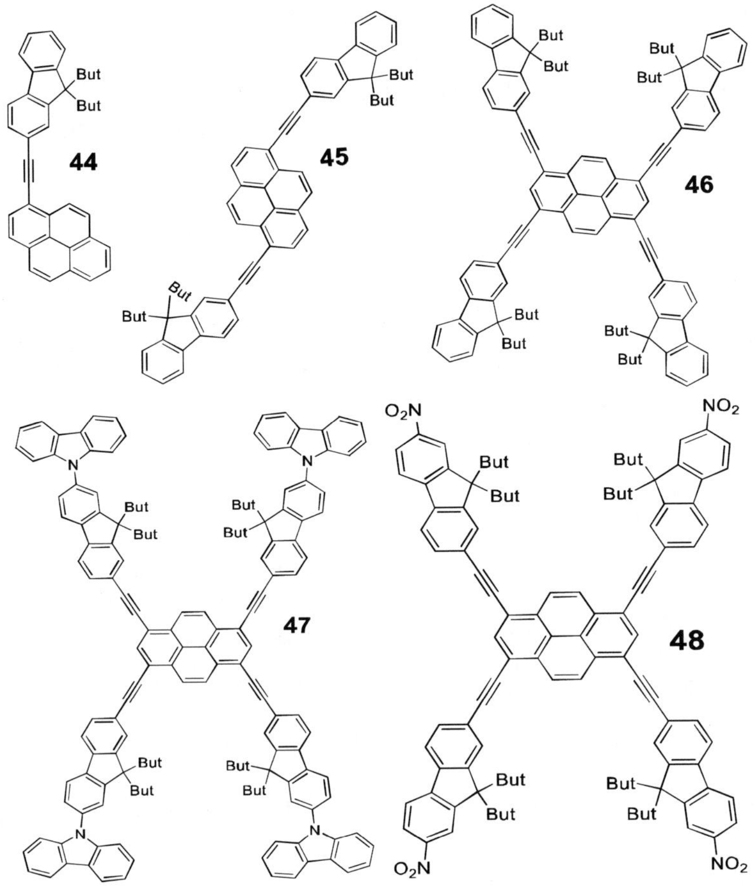

Scheme 14. Pyrene-acetylene linked to donor/acceptor compounds **44, 45, 46, 47** and **48** for two-photon absorption studies.

Pyrene-centred/core linked with donor or acceptor via acetylene linker compounds **44 - 48** (Lavanya Devi et al. 2015) were synthesized (Scheme 14) to study two-photon absorption properties and check their suitability in medical applications. Design informs that Donor -π- Acceptor (D- π -A) **44**; D-π-A-π-D **45, 46** and **47**, and A-π-D-π-A **48**. These type of D-A arrangements in a molecule provides extended π-conjugation, which leads to covering the molecular polarization distribution over the surface of the entire molecule and further this molecular polarization has a pivot role in the two-photon absorption property of the molecule. All compounds **44, 45, 46, 47** and **48** synthesized were taken for one photon and two-photon absorption (2PA) studies experimentally and theoretically to define the role of changes introduced in the structure. **44, 45, 46, 47** and **48** exhibited large Stokes shift, the high quantum yield of fluorescence and 2PA cross section values range 250 to 2500 GM. Suitable absorption, fluorescence, and thermal properties combined with photostability of these substances inform that they may be potential candidates for biological and medical applications. Particularly compound **47** showed the highest 2PA cross-section of 2460 GM compared to other compounds.

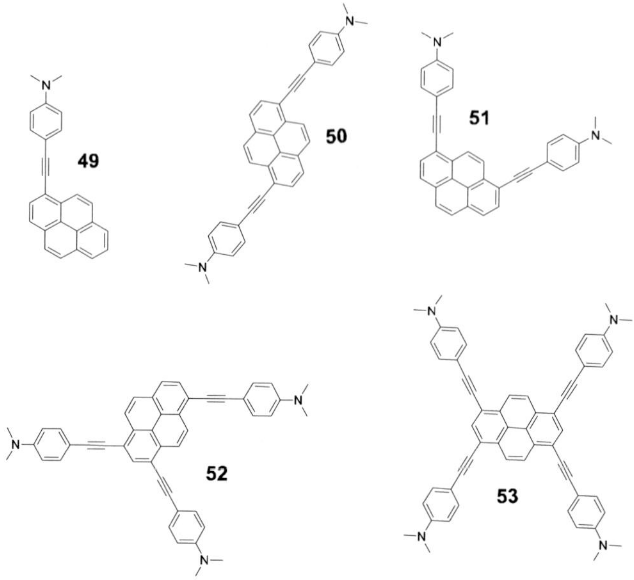

Scheme 15. Pyrene-Acetylene-N,N-Dimethylaminophenyl linked compounds **49, 50, 51, 52** and **53** for two-photon absorption studies.

Pyrene-centred, Pyrene-Acetylene-N, N-Dimethylaminophenyl linked Compounds **49, 50, 51, 52** and **53** (Hwan Myung Kim et al. 2008) were reported to study two-photon absorption (2PA) studies. All these compounds belong to the Donor- π -Acceptor type (D- π -A) design. The donor is the N, N-Dimethylamino-benzene, acetylene is the π link and pyrene is Acceptor. These molecules **49, 50, 51, 52** and **53** exhibited a gradual red shift in the one-photon absorption and fluorescence. The quantum yield of fluorescence of these molecules **49, 50, 51, 52** and **53** is found to be more than 90%. An increase in the red shift of fluorescence emission maxima was cited as an indication of the involvement of the intra-molecular charge transfer (ICT) excited state. The nonlinearity behaviour of these **49, 50, 51, 52** and **53** was confirmed by recording fluorescence intensity, which is proportional to the square of laser intensity employed. Compound **53** displayed the highest 2PA value of 1154 GM compared to other molecules. Authors indicate that pyrene can act as efficiently as anthracene as an emitter and is also good for further design of 2PA molecules.

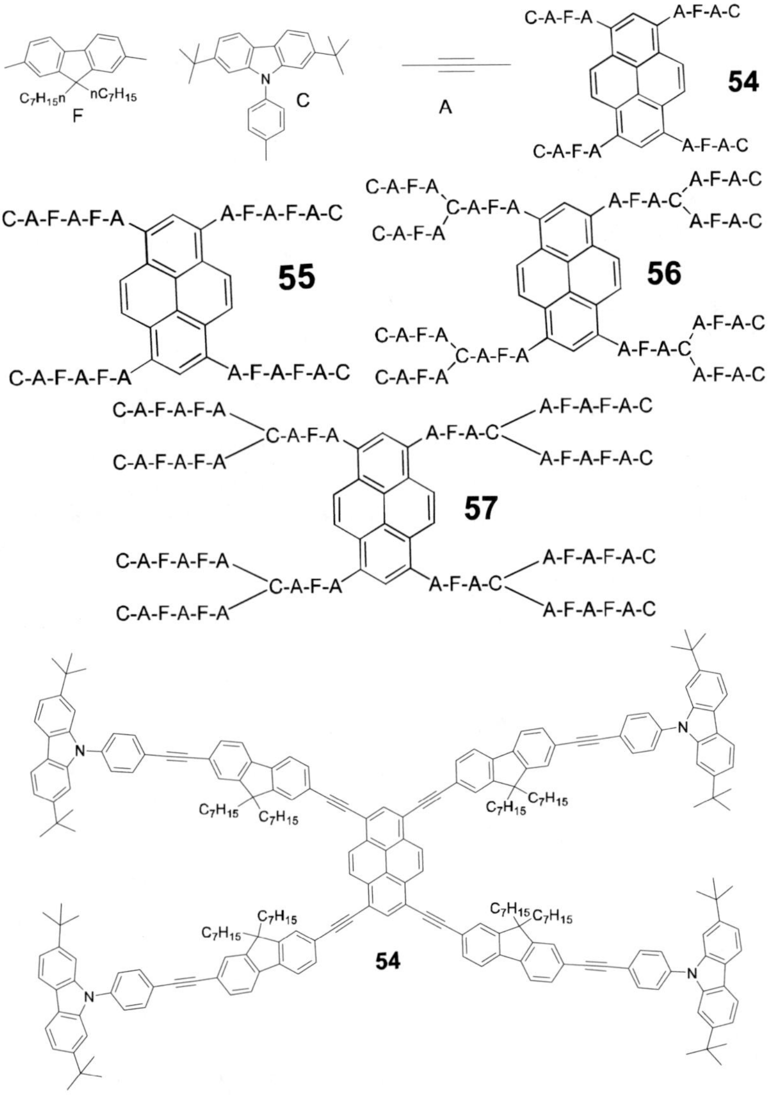

Scheme 16. Pyrene-Acetylene-Fluorene-ditBut-N-Phenylcarbazole combination of making dendrimers **54, 55, 56,** and **57** for two-photon absorption studies

Combination of Pyrene as the centre, attached with Acetylene – Fluorene – di-t-but-N-Phenylcarbazole used as design (Yan Wan et al. 2010) and synthesized dendrimers **54, 55, 56 and 57** (Scheme 16). All these molecules **54, 55, 56 and 57** were subjected to 2PA studies.

The Quantum yield of one photon-induced fluorescence of all the compounds determined is in the range of 0.49 to 0.52. All the compounds **54, 55, 56,** and **57** (Scheme 16) displayed 2PA capabilities and the values are in the range of 8,000 to 25,000 GM. Compound **57** is found to show ~25,000 GM value as the highest number. The enhancement in the 2PA numbers lies with the structure of the dendron, where each branch acts and contributes to in total. Possibly the energy transfer to the core may also be an important point of consideration. It will be interesting to say that the intramolecular charge transfer created by laser light absorption has got a large area to get distributed over it and this may help to improve the GM values.

Two different chalcones **58** and **59** (Scheme 17) **were** synthesized by using pyrene residue (Sheng-tao Shia et al. 2018) for two-photon absorption studies. The two chalcones **58** and **59** (Scheme 17) differ in the position or attachment of the carbonyl part of the molecule, which manifested in different and interesting 2PA properties and this triggers how a chemical structure can be related to its 2PA properties. One photon absorption spectrum of **59** is red-shifted relative to **58**. Chalcone **58** do not show any fluorescence and **59** exhibits excellent fluorescence. The chalcone **59**, as a small molecule, showed a remarkable 2PA cross-section of ~2400 GM and this is attributed to the "intermediate state resonance enhancement (ISRE)" phenomenon. The light-induced *cis-trans* (E – Z) isomerization of the α,β-unsaturated system is not discussed in this work on these two chalcones prepared. These chalcones **58,** and **59** carrying α,β-unsaturated system is prone to form a polar excited state upon light absorption (Rajgopal et al. 1997; Jayathirtha Rao 1994; Rajgopal et al. 1995). Authors claim that this two small chalcone molecules have good potential for application in optical data storage and optical limiting systems.

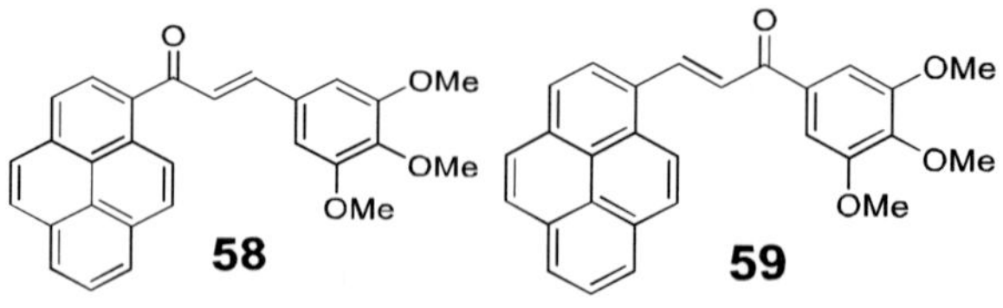

Scheme 17. Pyrene-Chalcones **58,** and **59** for two-photon absorption studies.

Pyrene-based chalcones **60** and **61** (Scheme 18) were synthesized (Ruipeng Niu et al. 2020) to study nonlinear optical properties and excited state dynamics. Transient absorption studies indicated that there are two singlet states, locally excited (LE) singlet formed by light absorption and followed by its (LE) internal conversion (2.5 n sec for **60** and 0.95 n sec for **61**), which leads to a nearby intramolecular charge transfer (ICT) excited state, applicable to both the molecules **60** and **61**. Pyrene acts as a donor in molecule **60** and acts as an acceptor in molecule **61**.

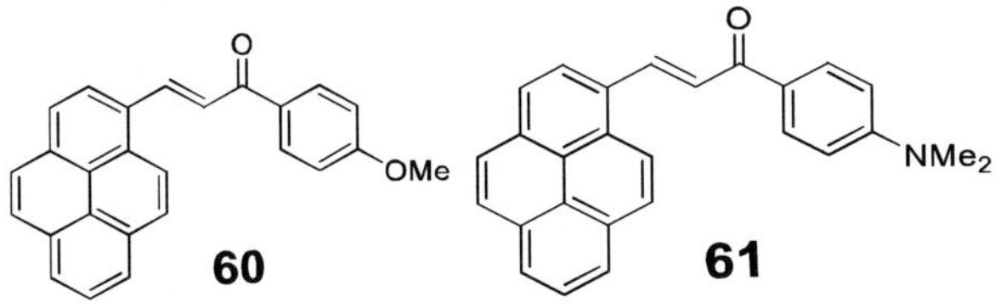

Scheme 18. Pyrene-Chalcones **60** and **61** for two-photon absorption studies.

Reverse saturable absorption was observed for both compounds **60** and **61**, upon conducting Z scan experiments. Observations inform that **60** has superior 2PA characteristics to **61**. But both the compounds showed good optical limiting performance indicating that these can be very good broad band optical limiting materials.

Theoretical investigations were conducted (Muhammad Usman Khan et al. 2022; Muhammad Khalid et al. 2012) involving pyrene and its attachment with D-π-D-π-A type combinations (Scheme 19) to evaluate NLO properties theoretically. Computational techniques provided data on (i) nonlinear optical properties; (ii) natural biding orbital; (iii) frontier molecular orbital; (iv) transition density matrics; and (v) UV-Visible absorption spectra. Frontier molecular orbital and natural binding orbitals combined information indicate that electron/charge flow is in the direction of the donor to acceptor leading to charge polarisation, which is required for a molecule to be NLO active. Further transition density metric data also informs that there is charge delocalisation within the molecule upon light absorption. UV-Visible absorption data simulated/derived from theoretical analysis (TD-DFT/M06/6-31G(d,p) level of theory) was used to understand HOMO, LUMO and other information. Results obtained from these computational studies provide that, the molecules studied are expected to display excellent NLO properties.

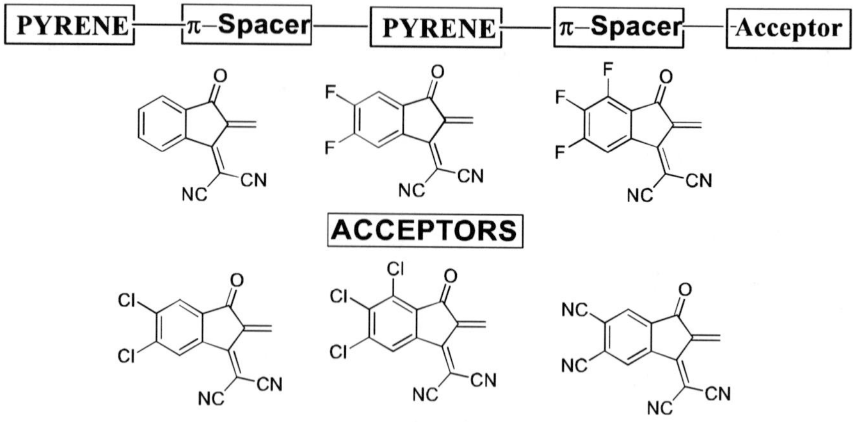

Scheme 19. Pyrene in combination with spacers and acceptors.

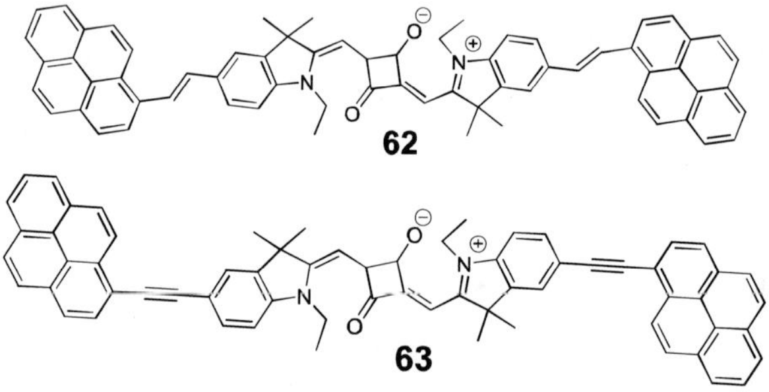

Scheme 20. Squaraine centred indolo-pyrene derivatives with spacers **62** and **63**.

Squaraine-centred indolo-pyrene derivatives **62** and **63** (Scheme 20) were synthesized (Alfonso et al. 2016) to study linear and nonlinear photophysical properties. Squaraine derivatives **62** and **63** (Scheme 20) differ in the π spacer, alkenyl (double bond) or alkynyl (triple bond). The π spacer, double bond or triple bond is expected to bring more planarity in the compounds **62** and **63** and improve intramolecular charge transfer interactions. Central squaraine (**62** and **63**) is the chromophore having high light absorption properties in the visible region. The two compounds **62** and **63** were subjected to absorption (1PA), photostability, fluorescence, quantum yields, solvatochromism, and fluorescence emission by varying the concentrations, two-photon absorption (2PA) and coupled with computational studies. Both compounds **62** and **63** displayed excellent photostability, high quantum yields of fluorescence and very good molar absorption coefficients. These compounds showed excimer formation at a higher levels of concentrations employed for fluorescence measurements. The 2PA data recorded for **62** and **63** exceeds 10,000 GM. Authors claim that putting the right spacers (alkene or alkyne) at the right place in the molecule provides enough planarity and thereby improves intramolecular charge transfer interactions leading to higher GM values noted (He et al. 2008).

Pyrene attached with pyridine – **64**, pyrene attached with pyridine via ethylenic link **65** and pyrene attached with pyridine via an alkyl (C_2H_4) link **66** were synthesized (Elena Lucenti et al. 2019) to evaluate linear and nonlinear optical properties. Compound **64**, where pyridine is directly attached to pyrene, in **65** pyrenes is attached to pyridine via ethylene link making them as push-pull type molecules (Scheme 21) and has the conjugative effect in operation, whereas compound **66** (Scheme 21) does not fall in this category. Crystal structures of **64** and **66** are also reported to understand the molecular arrangements in the solid state. UV-Vis absorption and fluorescence properties were evaluated in the normal solution phase, under the protonation state and after the deprotonation state using ammonia. Zinc complex of 64 and Cu complex of 66 were prepared and 1PA and fluorescence properties were reported. Electric Field Induced Second Harmonic (EFISH) measurements were determined for all the three **64, 65** and **66** compounds. The reading was found to be positive but it became negative after protonation. The negative reading for the protonated compound was reversed by introducing ammonia. The **64**-Zn complex and **66**-Cu complex displayed negative readings under EFISH measurements.

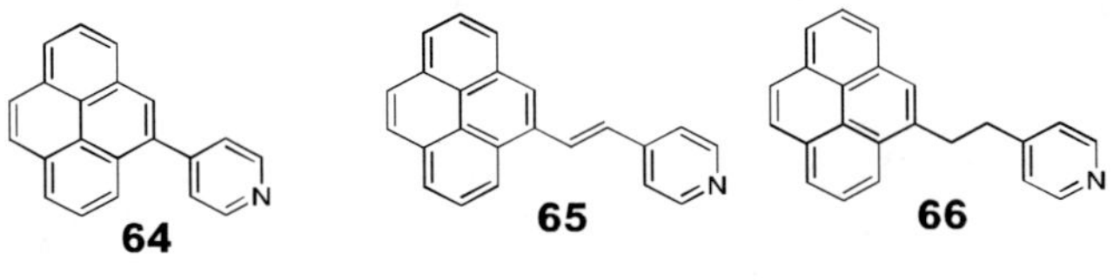

Scheme 21. Pyrene attached Pyridine compounds **64, 65** and **66**.

Conclusion

This chapter is on the role of pyrene in making various materials for practical applications. The survey is not exhaustive, but selective to bring the important approach to readers. The chapter starts with pyrene photo physical properties like absorption, fluorescence and fluorescence lifetime. A short introduction to OLEDs is arranged by covering important points. Several publications referred to highlight triplet energy transfer to pyrene singlet leading to improve electrochemiluminescence and this process helps in harvesting triplets formed from electrochemical means. The observed delayed fluorescence in OLED molecules referred, is cited as supporting the triplet to singlet energy transfer process. A brief narration on organic solar cells is covered to give an overall picture of various types of solar cells. Several examples cited inform that there is advancement in the research and organic solar cells with all small molecules have the future, besides perovskite solar cells. High efficiency of the organic solar cell can be obtained by careful design of small molecules. A quick understanding of NLO materials is attempted and also the possible applications of NLO materials required in future are also highlighted. Involvement of pyrene residue in designing NLO materials is brought to the reader by referring to a list of publications. Proper design of the molecule suitable for NLO activity is propagated. Attempts are made to convince the reader that these organic materials have a profound role to play in globally emerging technologies. The involvement of the polar excited states by means of light absorption is a necessary condition for both organic solar cells and organic NLO materials. The common phenomenon in these three types (OLEDs, Organic Solar Cells and NLO materials) of organic materials is the "highly polarized excited state" or "intramolecular charge transfer excited state" generated either by light absorption or by electrochemical means. Studies on the excited state dynamics are expected to through light on the mechanism of these complicated processes of applications.

Acknowledgments

VJR thank CSIR-New Delhi for Emeritus Scientist Honor and AcSIR-Ghaziabad for Emeritus Professor Honor.

References

Ballestas-Barrientos, Alfonso R., Woodward, Adam W., Moreshead, William V., Bondar, Mykhailo V. and Belfield, Kevin D. (2016) Synthesis and Linear and Nonlinear Photophysical Characterization of Two Symmetrical Pyrene-Terminated Squaraine Derivatives. *J Phys Chem C* 120, 7829-7838. DOI: 10.1021/acs.jpcc.6b00143.

Birks, J. B., Munro, I. H., Lumb, M. D. (1964) 'Excimer' fluorescence V. Influence of solvent viscosity and temperature. *Proc R Soc London A: Math Phys Sci* 280-289. https://doi.org/10.1098/rspa.1964.0146.

Bohne, C., Abuin, E. B. and Scaiano, J. C. (1990) Characterization of the Triplet-Triplet Annihilation Process of Pyrene and Several Derivatives under Laser Excitation. *J Am Chem Soc* 112, 4226-4231. doi:10.1021/ja00167a018.

Brunner, Klemens, Addy van Dijken, Herbert Bo"rner, Jolanda Bastiaansen, J.A.M. Nicole Kiggen, M. M. and Bea Langeveld, M. W. (2004) Carbazole Compounds as Host Materials for Triplet Emitters in Organic Light-Emitting Diodes: Tuning the HOMO Level without Influencing the Triplet Energy in Small Molecules. *J Am Chem Soc* 126, 6035-6042. https://doi.org/10.1021/ja049883a.

Chen, G., Qiu, Z. Tan, J.-H Chen, W.-C Zhou, P. Xing, L. Ji, S. Qin, Y. Zhao, Z. Huo, Y. (2021) Deep-blue organic light-emitting diodes based on push-pull π-extended imidazole-fluorene Hybrids. *Dyes Pigments* 184, 108754. doi: 10.1016/j.dyepig. 2020.108754.

Chen, Guowei, Zhipeng Qiu, Ji-Hua Tan, Wen-Cheng Chen, Peiqi Zhou, Longjiang Xing, Shaomin Ji, Yanlin Qin, Zujin Zhao, Yanping Huo (2021) Deep-blue organic light-emitting Diodes based on push-pull π-extended imidazole-fluorene hybrids. *Dyes and Pigments* 184, 108754. https://doi.org/10.1016/j.dyepig.2020.108754.

Chercka, D., Yoo, S.J., Baumgarten, M., Kim, J.-J., Müllen, K. (2014) Pyrene based materials for exceptionally deep blue OLEDs, *J. Mater. Chem. C.* 2, 9083–9086, doi: 10.1039/C4TC01801J.

Chidirala, Swetha, Hidayath Ulla, Anusha Valaboju, Raveendra Kiran, M., Maneesha Esther Mohanty, Satyanarayan, M. N., Umesh, G. Kotamarthi Bhanuprakash, Vaidya Jayathirtha Rao (2016) Pyrene-Oxadiazoles for Organic Light Emitting Diodes: Triplet to Singlet Energy Transfer and Role of Hole-injection/Hole-blocking Materials. *J Org Chem* 81, 603-614. DOI: 10.1021/acs.joc.5b02423.

Clar, E., Gwe-Vuillbme, J. F., Mccallum, S. and McPherson, I. A. (1963) Annellation Effects In the Pyrene Series and the Classification of Absorption Spectra. *Tetrahedron* 19, 2185-2197. https://doi.org/10.1016/0040-4020(63)85034-6.

Clar, E., Schmidt, W. (1976) Correlations between photoelectron and phosphorescence spectra of polycyclic hydrocarbons. *Tetrahedron,* 32, 2563-2566. https://doi.org/10. 1016/00404020(76)88027-1.

Crawford, A. G., Dwyer, A. D., Liu, Z., Steffen, A., Beeby, A., Palsson, L., Tozer, D. J. and Marder, T. B. A. (2011) Experimental and Theoretical Studies of the Photophysical Properties of 2- and 2,7-Functionalized Pyrene Derivatives *J Am Chem Soc* 133, 13349. dx.doi.org/10.1021/ja2006862.

Cui, Lin-Song, Shi-Bin Ruan, Fatima Bencheikh, Ryo Nagata, Lei Zhang, Ko Inada, Hajime Nakanotani, Liang-Sheng Liao & Chihaya Adachi (2017) Long-lived efficient delayed fluorescence organic light-emitting diodes using n-type hosts. *Nature Communications* 8, 2250. DOI: 10.1038/s41467-017-02419-x.

Delouis, J-F., Delaire, J-A. and Ivanoff, N. (1979) Pyrene fluorescence quenching and triplet- state formation in the presence of DABCO (1,4-Diazabicyclo[2-2-2]octane): A laser photolysis study. *Chem Phys Lett* 61, 343-346. DOI: 10.1016/0009-2614(79) 80659-4.

Devi, Lavanya, C. Kada Yesudas, Nikolay Makarov, S. Vaidya Jayathirtha Rao, Kotamarthy Bhanuprakash and Joseph William Perry (2015) Fluorenylethynylpyrene Derivatives with Strong Two-Photon Absorption: Influence of Substituents on Optical Properties. *J Mater Chem C* 3, 3730-3744. https://doi.org/10.1039/C4TC02896A.

Feng, Tang, Kaile Wu, Zhijie Zhou, Guo Wang, Bin Zhao, and Songting Tan (2019) Alkynyl- Functionalized Pyrene-Cored Perylene Diimide Electron Acceptors for Efficient Non-fullerene Organic Solar Cells. *ACS Appl Energy Mater* 2, 3918-3926. DOI: 10.1021/acsaem.9b00611.

Figueira-Duarte, T. M. and Klaus Mullen (2011) Pyrene Based Materials for Organic Electronics. *Chem Rev*, 111, 7260–7314. dx.doi.org/10.1021/cr100428a

Foster, T. Kasper, K. (1954) Ein Konzentrationsumschlag der Fluoreszenz [A change in concentration of the fluorescence]. *Z Physik Chem* 1, 275. https://doi.org/10.1524/zpch.1954.1.5_6.275.

Ge, Y. Wen, Y. Liu, H. Lu, T. Yu, Y. Zhang, X. Li, B. Zhang, S.-T. Li, W. Yang, B. (2020) A Key stacking factor for the effective formation of pyrene excimer in crystals: degree of π–π Overlap. *J Mater Chem C* 8, 11830–11838. doi: 10.1039/ D0TC02562C.

Godumala, Mallesham, Chidirala Swetha, Surukonti Niveditha, Maneesha Esther Mohanty, Nanubolu Jagadeesh Babu, Arunandan Kumar, Kotamarthi Bhanuprakash, and Vaidya Jayathirtha Rao (2015) Phosphine Oxide Functionalized Pyrenes as Efficient Blue Light Emitting Multifunctional Materials for organic Ligh Emitting Diodes. *J Mater Chem C* 3, 1208-1224. DOI: 10.1039/c4tc01753f.

Gopal, Raj, V. Jayathirtha Rao, V. Saroja, G. Samanta, A. (1997) Photophysical behaviour of some pyrenylethylene derivatives and its implication on trans cis photo-isomerisation reactions. *Chem Phys Lett* 270, 593-598. https://doi.org/10.1016/S0009-2614(97)00394-1.

Goushi, Kenichiand Chihaya Adachi (2012) Efficient organic light-emitting diodes through up- conversion from triplet to singlet excited states of exciplexes. *Appl Phys Lett* 101, 023306. doi: 10.1063/1.4737006.

Hajime, Maeda, Tomohiro Maeda, Kazuhiko Mizuno, Kazuhisa Fujimoto, Hisao Shimizu, and Masahiko Inouye (2006) Alkynylpyrenes as Improved Pyrene-Based Bio-molecular Probes with The Advantages of High Fluorescence Quantum Yields and Long Absorption/Emission Wavelengths. *Chem Eur J* 12, 824 – 831. DOI: 10.1002/chem.200500638.

Hammarstrom, P., Kalman, B. Jonsson, B. H. Carlsson, U. (1997) Pyrene excimer fluorescence as a proximity probe for investigation of residual structure in the unfolded state of human carbonic anhydrase II. *FEBS Lett* 420, 63. https://doi.org/10.1016/S0014-5793(97)01488-9.

Hautala, R. R., Schore N. E and. Turro N. J. (1973) Novel fluorescent probe. Use of time-correlated fluorescence to explore the properties of micelle-forming detergent. *J Am Chem Soc* 95, 5508-5514. https://doi.org/10.1021/ja00798a013.

He, G. S., Tan, L.-S., Zheng, Q., Prasad, P. N. (2008) Multiphoton Absorbing Materials: Molecular Designs, Characterizations, and Applications. *Chem Rev* 108, 1245−1330. DOI: 10.1021/cr050054x.

Hofstraat, J. W. Engelsma, M. Cofino, M. W. P. Hoornweg, G. Ph. Gooijer, C. and Velthorst, N. H. (1984) Highly Resolved Fluorescence Spectrometry of Pyrene on a

Thin Layer Chromatography Plate. *Analytica Chimica Acta* 159, 359-363. https://doi.org/10.1016/S0003-2670(00)84312-3.

Idris, Muazzam, Caleb Coburn, Tyler Fleetham, JoAnna Milam-Guerrero, Peter Djurovich, I, Stephen Forrest, R. and MarkThompson, E. (2019) Phenanthro [9,10-d]triazole and imidazole derivatives: high triplet energy host materials for blue phosphorescent organic light emitting devices. *Mater Horiz* 6, 1179—1186. DOI: 10.1039/c9mh 00195f.

Iu, K.-K and Thomas, J. K. (1990) Photophysical Properties of Pyrene in Zeolites. 2. Effects of Coadsorbed Water. *Langmuir* 6, 471-478. https://doi.org/10.1021/la00092a029.

Jayathirtha Rao, Vaidya (1994) Regioselective photoisomerization of dienones: role of polar excited state. *J Photochem Photobiol A Chem* 83, 211-215. https://doi.org/10.1016/1010-6030(94)03833-3.

Jiang, M., Wang, X., Xu, Q., Li, M., Niu, D, Sun, X., Wang, F., Li, Z., Zhang, D. (2018) High-speed electro-optic switch based on nonlinear polymer-clad waveguide incorporated with quasi-in- plane coplanar waveguide electrodes. *Opt Mater* 75, 26–30. https://doi.org/10.1016/j.optmat.2017.10.020.

Jung, Mina, Jaehyun Lee, Hyocheol Jung, Seokwoo Kang, Atsushi Wakamiya, Jongwook Park (2018) Highly efficient pyrene blue emitters for OLEDs based on substitution position effect. *Dyes and Pigments* 158, 42-49. https://doi.org/10.1016/j.dyepig.2018.05.024.

Khalid, Muhammad, Hafiza Munnazza Lodhi, Muhammad Usman Khan and Muhammad Imran (2012) Structural parameter-modulated nonlinear optical amplitude of acceptor–p–D–p–donor Configured pyrene derivatives: a DFT approach. *RSC Adv* 11, 14237–14250. DOI: 10.1039/d1ra00876e.

Kim, Ji-Hoon, Hee Un Kim, In-Nam Kang, Sang Kyu Lee, Sang-Jin Moon, Won Suk Shin, and Do-Hoon Hwang (2012) Incorporation of Pyrene Units to Improve Hole Mobility in Conjugated Polymers for Organic Solar Cells. *Macromolecules* 45, 8628–8638. dx.doi.org/10.1021/ma301877q.

Kumar, Krishan, Kiran Kishore Kesavan, Diksha Thakur, Subrata Banik, Jayachandran Jayakumar, Chien-Hong Cheng, Jwo-Huei Jou, and Subrata Ghosh (2021) Functional Pyrene–Pyridine-Integrated Hole-Transporting Materials for Solution-Processed OLEDs with Reduced Efficiency Roll-Off. *ACS Omega* 6, 10515–10526. https://doi.org/10.1021/acsomega.0c04080.

Lambert, C., C. Ehbets, J. Rausch, D. and Steeger, M. (2012) Charge-Transfer Interactions in a Multichromophoric Hexaarylbenzene Containing Pyrene and Triarylamines. *J Org Chem* 77, 6147. https://doi.org/10.1021/jo300924x.

Langelaar, J., Rettschnick, R., P. H. and Hoijtink, G. J. (1971) Studies on Triplet Radiative Lifetimes, Phosphorescence and Delayed Fluorescence Yields of Aromatic Hydrocarbons in Liquid Solutions. *The J Chem Phys* 54, 1-7. https://doi.org/10.1063/1.1674576.

Laurent, A. (1837). Suite De Recherches Diverses De Chimie Organique [Suite Of Various Researches Of Organic Chemistry]. *Ann Chim Phys Ser* 66, 326–337.

Lewis, F. D. Zhang, Y. F. Letsinger, R. L. (1997) Bispyrenyl Excimer Fluorescence: A Sensitive Oligonucleotide Probe. *J Am Chem Soc* 119, 5451. https://doi.org/10.1021/ja9641214.

Liu, X. Iu, K.-K and Thomas, J. K. (1989) Photophysical Properties of Pyrene in Zeolites. *J Phys Chem* 93, 4120-4128. https://doi.org/10.1021/j100347a049.

Liu, Y. Bai, Q. Li, J. Zhang, S. Zhang, C. Lu, F. Yang, B. Lu, P. (2016) Efficient pyrene-imidazole derivatives for organic light-emitting diodes. *RSC Adv* 6, 17239–17245. doi: 10.1039/C5RA25424H.

Lucenti, Elena, Alessandra Forni, Daniele Marinotto, Andrea Previtali, Stefania Righetto and Elena Cariati (2019) Tuning the Linear and Nonlinear Optical Properties of Pyrene-Pyridine Chromophores by Protonation and Complexation to d10 Metal Centers. *Inorganics* 7, 38-47. doi:10.3390/inorganics7030038.

Marpaung, D., Yao J., Capmany J. (2019) Integrated microwave photonics. *Nat Photon* 13, 80–90. https://doi.org/10.1038/s41566-018-0310-5.

Mun, Jeong-Wook, Illhun Cho, Donggu Lee, Won Sik Yoon, Oh Kyu Kwon, Changhee Lee, Soo Young Park (2013) Acetylene-bridged D–A–D type small molecule comprising pyrene and diketopyrrolopyrrole for high efficiency organic solar cells. *Org Electronics* 14, 2341-2347. http://dx.doi.org/10.1016/j.orgel.2013.05.035.

Myung Kim, Hwan, Yeon Ok Lee, Chang Su Lim, Jong Seung Kim, and Bong Rae Cho (2008) Two-Photon Absorption Properties of Alkynyl-Conjugated Pyrene Derivatives. *J Org Chem* 73, 5127-5130. DOI: 10.1021/jo800363v.

Niu, Ruipeng, Yuxiao Wang, Xingzhi Wu, Shuang Chen, Xueru Zhang, and Yinglin Song (2020) D–π–A-Type Pyrene Derivatives with Different Push−Pull Properties: Broadband Absorption Response and Transient Dynamic Analysis. *J Phys Chem C* 124, 5345−5352. https://dx.doi.org/10.1021/acs.jpcc.9b11667.

Ollmann, M. Schwarzmann, G. Sandhoff, K. Galla, H. J. (1987) Pyrene-labeled gangliosides: micelle formation in aqueous solution, lateral diffusion, and thermostropic behavior in phosphatidylcholine bilayers. *Biochemistry* 26, 5943. https://doi.org/10.1021/bi00392a055.

Ottonelli, Massimo, Matteo Piccardo, Daniele Duce, Sergio Thea, and Giovanna Dellepiane (2012) Tuning the Photophysical Properties of Pyrene-Based Systems: A Theoretical Study. *J Phys Chem A* 116, 611–630. dx.doi.org/10.1021/jp2084764.

Pap, E. H. W., Hanicak, A., Vanhoek, A., Wirtz, K. W. A., Visser, A. J. W. G. (1995) Quantitative Analysis of Lipid-Lipid and Lipid-Protein Interactions in Membranes by Use of Pyrene-Labeled Phosphoinositides. *Biochemistry* 34, 9118-9125. https://doi.org/10.1021/bi00028a022.

Paris, P. L. Langenhan, J. M. Kool, E. T. (1998) Probing DNA sequences in solution with a monomer-excimer fluorescence color change. *Nucleic Acids Res* 26, 3789. https://doi.org/10.1093/nar/26.16.3789.

Pati, A. K. Gharpure, S. J. and Mishra, A. K. (2015) On the photophysics of butadiyne bridged pyrene–phenyl molecular conjugates: multiple emissive pathways through locally excited, intramolecular charge transfer and excimer states. *Faraday Discus* 177, 213. https://doi.org/10.1039/C4FD00170B.

Pokhrel, M. R., Bossmann, S. H. (2000) Synthesis, Characterization, and First Application of High Molecular Weight Polyacrylic Acid Derivatives Possessing Perfluorinated Side Chains and Chemically Linked Pyrene Labels. *J Phys Chem B* 104, 2215-2223. https://doi.org/10.1021/jp9917190.

Pope, M. Kallman, H. P. and Magnante, P. (1963) Electroluminescence in Organic Crystals. *J Chem Phys* 38, 2042-2043. https://doi.org/10.1063/1.1733929.

Prabhakar, Chetti, Gunturu Krishna Chaitanya, Sanyasi Sitha, Kotamarthi Bhanuprakash and Vaidya Jayathirtha Rao (2005) Role of the Oxyallyl Substructure in the Near Infrared (NIR) Absorption in Symmetrical Dye Derivatives: A Computational Study. *J Phys Chem A* 109, 2614–2622. https://doi.org/10.1021/jp044954d.

Raj, Gopal, V. Mahipal Reddy, A. and Vaidya Jayathirtha Rao (1995) Wavelength Dependent Trans to Cis and Quantum Chain Isomerizations of Anthrylethylene Derivatives. *J Org Chem* 60, 7966–7973. https://doi.org/10.1021/jo00129a043.

Raj, Mohan, A. Gaurav Sharma, Rajeev Prabhakar and Ramamurthy, V. (2019) Room-Temperature Phosphorescence from Encapsulated Pyrene Induced by Xenon. *J. Phys. Chem. A* 123, 9123−9131. DOI: 10.1021/acs.jpca.9b08354.

Rajgopal, V. Anugula Mahipal Reddy and Vaidya Jayathirtha Rao (1995) Wavelength Dependent Trans to Cis and Quantum Chain Isomerizations of Anthrylethylene Derivatives. *J Org Chem* 60, 7966–7973. https://doi.org/10.1021/jo00129a043.

Sahoo, D. Weers, P. M. M. Ryan, R. O. Narayanaswami, V. (2002) Lipid-triggered conformational switch of apolipophorin III helix bundle to an extended helix organization. *J Mol Biol* 321, 201. DOI: 10.1016/s0022-2836(02)00618-6.

Sahoo, D., Narayanaswami, V., Kay, C. M., Ryan, R. O. (2000) Pyrene Excimer Fluorescence: A Spatially Sensitive Probe to Monitor Lipid-Induced Helical Rearrangement of Apolipophorin III. *Biochemistry* 39, 6594. https://doi.org/10.1021/bi992609m.

Shia, Sheng-tao, Yu Fangb, Jun-yi Yanga, Yan-bing Hanc, Ying-lin Songa, (2018) The remarkable enhancement of two-photon absorption in pyrene basedchalcone derivatives. *Optical Materials* 86, 331–337. https://doi.org/10.1016/j.optmat.2018.10.026.

Shungang Liu, Wenyan Su, Xianshao Zou, Xiaoyan Du, Jiamin Cao, Nong Wang, b Xingxing Shen, Xinjian Geng, Zilong Tang, Arkady Yartsev, Maojie Zhang, Wolfgang Gruber, Tobias Unruh, Ning Li, Donghong Yu, Christoph Brabece and Ergang Wang (2020) The role of connectivity in significant bandgap narrowing for fused-pyrene based non-fullerene acceptors toward high-efficiency organic solar cells. *J Mater Chem A* 8, 5995-6003. DOI: 10.1039/d0ta00520g.

Song, X. D. Swanson, B. I. (1999) Rational Design of an Optical Sensing System for Multivalent Proteins. *Langmuir* 15, 4710-4712. https://doi.org/10.1021/la980758k.

Soon, Ok Jeon and Jun Yeob Lee (2012) Phosphine oxide derivatives for organic light emitting Diodes. *J Mater Chem* 22, 4223-4233. DOI: 10.1039/c1jm14832j.

Soultati, Anastasia, Apostolis Verykios, Stylianos Panagiotakis, Konstantina-Kalliopi Armadorou, Muhammad Irfan Haider, Andreas Kaltzoglou, Charalampos Drivas, Azhar Fakharuddin, Xichang Bao, Chunming Yang, Abd. Rashid bin Mohd Yusoff, Evangelos K. Evangelou, Ioannis Petsalakis, Stella Kennou, Polycarpos Falaras, Konstantina Yannakopoulou, George Pistolis, Panagiotis Argitis, and Maria Vasilopoulou (2020) Suppressing the Photocatalytic Activity of Zinc Oxide Electron-Transport Layer in Nonfullerene Organic Solar Cells with a Pyrene-Bodipy Interlayer. *ACS Applied Materials & Interfaces* 12, 21961-21973. https://dx.doi.org/10.1021/acsami.0c03147.

Sylvinson, Daniel, M. R. Chen, H. F. Martin, L. M. Saris, P. J. G. and Thompson, M. E. (2019) Rapid Multiscale Computational Screening for OLED Host Materials. *ACS Appl Mater Interfaces* 11, 5276–5288. https://doi.org/10.1021/acsami.8b16225.

Tang, C. W. and VanSlyke, S. A. (1987) Organic electroluminescent diodes. *Appl Phys Lett* 51, 913-915. https://doi.org/10.1063/1.98799.

Tao, S. L., Peng, Z. K., Zhang, X. H., Wang, P. F., Lee, C.-S., & Lee, S.-T. (2005) Highly Efficient Non-Doped Blue Organic Light-Emitting Diodes Based on Fluorene Derivatives with High Thermal Stability. *Advanced Functional Materials*, 15,1716-1721. doi:10.1002/adfm.200500067.

Tao, Silu, Yechun Zhou, Chun-Sing Lee, Xiaohong Zhang, and Shuit-Tong Lee (2010) High-Efficiency Nondoped Deep-Blue-Emitting Organic Electroluminescent Device. *Chem Materials* 22, 2138-2141. DOI:10.1021/cm100100w.

Technical development roadmap of China's optoelectronic device industry (2018- 2022), *Electronic intellectual property* (01). 2018. p. 9.

Templer, R. H. Castle, S. J. Curran, A. R. Rumbles, G. Klug, D. R. (1998) Sensing isothermal changes in the lateral pressure in model membranes using di-pyrenyl phosphatidylcholine *Faraday Discuss* 11, 41-53. https://doi.org/10.1039/A806472E.

Tong, G. Lawlor, J. M. Tregear, G. W. Haralambidis, J. (1995) Oligonucleotide-Polyamide Hybrid Molecules Containing Multiple Pyrene Residues Exhibit Significant Excimer Fluorescence. *J Am Chem Soc* 117, 12151. https://doi.org/10.1021/ja00154a015.

Umasankar, Gorakala, Manohar Reddy Busireddy, Bhanuprakash Kotamarthi, Galla Karunakar, V. and Vaidya Jayathirtha Rao (2020) Pyrene appended terpyridine derivatives as electrochemiluminescence material for OLEDs: Characterization of photo-physical, thermal and electrochemical properties. *Ind J Chem B* 59B, 1564-1574.

Umashankar, Gorakal, Hidayat Ulla, Chakali Madhu, Gontu Ramanjeneya Reddy, Balaiah Shanigaram, Jagadeesh babu nanubolu, Bhanuprakash Kotamarthi, Galla Karunakar, V. Satyanarayan Nagarajan, M. and Vaidya Jayathirtha Rao (2021) Imidazole-Pyrene Hybrid Luminescent Materials for Organic Light Emitting Diodes: Synthesis, Characterization & Electrochemiluminescent Properties. *Journal of Moleculr Structure* 1236, 130306. https://doi.org/10.1016/j.molstruc.2021.130306.

Usman Khan, Muhammad, Muhammad Khalid, Rasheed Ahmad Khera, Muhammad Nadeem Akhta, Amna Abbas, Muhammad Fayyaz ur Rehman, Ataualpa Albert Carmo Braga, Mohammed Mujahid Alam, Muhammad Imra, Yao Wang, Changrui Lu (2022) Influence of acceptor tethering on the performance of nonlinear optical properties for pyrene-based materials with A-p-D-p-D architecture. *Arabian Journal of Chemistry* 15, 103673. https://doi.org/10.1016/j.arabjc.2021.103673.

Wan, Yan, Linyin Yan, Zujin Zhao, Xiaonan Ma, Qianjin Guo, Mingli Jia, Ping Lu, Gabriel Ramos-Ortiz, Jose' Luis Maldonado, Mario Rodrı´guez, and Andong Xia, (2010) Gigantic Two-Photon Absorption Cross Sections and Strong Two-Photon Excited Fluorescence in Pyrene Core Dendrimers with Fluorene/Carbazole as Dendrons and Acetylene as Linkages. *J. Phys. Chem. B* 114, 11737–11745. https://doi.org/10.1021/jp104868j.

Winnik, F. M. (1993) Photophysics of preassociated pyrenes in aqueous polymer solutions and in other organized media. *Chem Rev,* 93, 587. https://doi.org/10.1021/cr00018a001.

Yamana, K. Iwai, T. Ohtani, Y. Sato, S. Nakamura, M. Nakano, H. (2002) Bis-Pyrene-Labeled Oligonucleotides: Sequence Specificity of Excimer and Monomer Fluorescence Changes upon Hybridization with DNA. *Bioconjugate Chem* 13, 1266. https://doi.org/10.1021/bc025530u.

Yamana, K. Takei, M. Nakano, H. (1997) Synthesis of oligonucleotide derivatives containing pyrene labeled glycerol linkers: Enhanced excimer fluorescence on binding to a complementary DNA sequence. *Tetrahedron Lett* 38, 6051. https://doi.org/10.1016/S0040-4039(97)01358-0.

Biographical Sketch

Vaidya Jayathirtha Rao

Affiliation: CSIR-Indian Institute of Chemical Technology, Uppal Road Tarnaka, Hyderabad-500007, Telangana, India and AcSIR-Ghaziabad

Education:
BSc –1976 Gadwal-509125, Hometown;
MSC-1978 Univ.Sci.College, Osmania Univ. Hyderabad-500007;
PhD –1983 Indian Institute of Science, Bengaluru-560012

Business Address: Fluor-Agro Chemicals Department, CSIR-Indian Institute of Chemical Technology, Uppal Road Tarnaka, Hyderabad-500007, Telangana, India and AcSIR-Ghaziabad

Research and Professional Experience: >40 years; AvHFellow; FRSC; Member ACS since 2009

Professional Appointments: Chief Scientist; Head Department; Director Grade Scientist; Professor; Jury in a Court; Advisor to Research Institute;

Honors:
Recent Recognitions:
  (1) Telangana District Scientist – 2017;
  (2) Inspirational Member Award by Royal Society of Chemistry – London – 2019;

(3) Fellow Chartered Chemist – RSC –London – 2020;
(4) Member Appellate Authority (Tribunal Court) Telangana State Pollution Control-2021;
(5) Top 2% Researcher in India (Medicinal and Biomolecular Chemistry) – Elsevier & Stanford Univ. Analysis – 2020;
(6) Top 2% Researcher in India (Medicinal and Biomolecular Chemistry) – Elsevier & Stanford Univ. Analysis – 2021.

Publications from the Last 3 Years: >20. Google Scholar Citations of Jayathirtha Rao Vaidya

Chapter 3

The Presence of Micro and Nanoplastics Influencing Pyrene Toxicity

Hillary Henao-Toro[1], Edwin Chica[1], PhD and Ainhoa Rubio-Clemente[1,2,*], PhD

[1]Grupo de Investigación Energía Alternativa (GEA), Facultad de Ingeniería, Universidad de Antioquia, Medellín, Colombia
[2]Escuela Ambiental, Facultad de Ingeniería, Universidad de Antioquia, Medellín, Colombia

Abstract

Pyrene is a polycyclic aromatic hydrocarbon with carcinogenic, mutagenic and teratogenic potential. In turn, micro and nanoplastics have been recently known as contaminants of emerging concern, whose detection and quantification in hydric resources is increasing every day. Due to the physicochemical properties of pyrene, especially the low solubility in water and the high affinity for particulate matter, this substance has a tendency to be sorbed into solid materials such as micro and nanoplastics, which are extensively distributed worldwide. In this work, the influence of the presence of micro and nanoplastics in particularly aquatic ecosystems on pyrene toxicity is analyzed. It was evidenced that even though micro and nanoplastics delayed the appearance of pyrene potential detrimental effects on biota, a synergic effect can occur, leading to drastic consequences on aquatic fauna and flora biodiversity. In this regard, the co-existence of pyrene and micro

---

[*] Corresponding Author's Email: ainhoa.rubioc@udea.edu.co.

In: Pyrene: Chemistry, Properties and Uses
Editor: Charles R. Howe
ISBN: 979-8-88697-670-0
© 2023 Nova Science Publishers, Inc.

and nanoplastic particles in the environment need to be urgently addressed so that the biota risks associated are minimized.

Keywords: water pollution, emerging pollutants, ecotoxicity, monitoring, polycyclic aromatic hydrocarbons

Introduction

Water is essential for life (Rubio-Clemente et al., 2019). It is a resource of great importance for ecosystems, as well as for human society, due to the ecosystem services it provides (Aznar-Sánchez et al., 2019). However, nowadays, there are more and more pollutants found in water due not only to the exponential growth of the population but also the disproportionate increase in industrial activity, urban expansion and intensification of agricultural practices (Feng et al., 2018), among other activities. The incorrect treatment, or even the non-existence, that is done to the polluted water prior to its discharge, increases the pollutant load that they receive, contributing to a rapid growth in the loss of aquatic ecosystems and their biodiversity (Paredes et al., 2019). This is ascribed to 40% reduction in the quality of the water resource by 2030 (WWAP, 2015; Salehi et al., 2022).

There are a large number of pollutants that constantly threaten the quality of water sources and endanger the aquatic ecosystems. Among them, plastics are especially relevant. Plastic has now become an indispensable material for agriculture, livestock and industry in general. Likewise, plastic is commonly used in daily life (Gao et al., 2022), leading to a large amount of organic waste that is difficult to be treated. In fact, by 2018, the annual production of plastic reached approximately 359 million Tons (PlasticsEurope, 2019). The problem caused by plastics in the environment is fundamentally attributed to their persistent nature, degradation and mineralization. It should be noted that the plastics contained in the environment can be gradually fragmented into microplastics (1 μm < size < 5 mm) and nanoplastics (< 1000 nm) (Monikh et al., 2022), due to synergistic physicochemical and biological processes (Huang et al., 2021). In this regard, the presence of emerging contaminants (ECs), such as micro and nanoplastics, in the environment increases (López et al., 2022). Furthermore, a rise in the toxicity associated with the existence of this type of materials in the environment is also observed; because of their size, micro and nanoplastics are easily absorbed into the aquatic environment by the fauna and flora inhabiting it (Sun et al., 2021). It is noteworthy that micro and nanoplastics are ingested, either directly or indirectly, by aquatic organisms present at different trophic levels, representing a significant alteration in food chains and triggering several negative impacts on the vital activities of organisms (Huang et al., 2021).

On the other hand, micro and nanoplastic particles can serve as transport vehicles for other contaminants and even pathogenic organisms. Physico-chemical properties and weathering processes in nature affect the capacity of micro and nanoplastics to adsorb/desorb chemicals in the environment, as well as the extent to which they release toxic chemicals into the environment (Huang et al., 2021). For example, large amounts of micro-nanoplastics found in environmental matrices can easily adsorb pollutants, such as polychlorinated biphenyls, organochlorine pesticides, antibiotics and heavy metals (Anbumani et al., 2018), being polycyclic aromatic hydrocarbons (PAHs) pollutants of special relevance.

PAHs are persistent organic pollutants (Wang et al., 2022) that are ubiquitously found in the environment. They are generated naturally through volcanic explosions and forest fires, and anthropically in the pyrolysis processes of organic matter and fossil fuels (Zhang et al., 2019; Sun et al., 2021; Rubio-Clemente et al., 2014). PAHs are detected in water masses and sediments, air, soils, and animal and plant tissues (Zhang et al., 2019), and represent a problem for public health and ecosystems, being a topic of interest in researches related to their presence and distribution. Additionally, the existence of these pollutants in water exhibits a high risk, particularly those with a carcinogenic, mutagenic and teratogenic character, due to their direct implications in the deterioration of human health (Wang et al., 2018). Some of the direct risks associated with the exposure to PAHs are fertilization failure, limitation of the embryonic development and oxidative damage, among others (Wang et al., 2022). Due to the high levels of toxicity ascribed to this type of substance, the United States Environmental Protection Agency (EPA) has included 16 PAHs in the list of priority pollutants, among which pyrene is named (Yan et al., 2004).

On the other hand, due to the capacity of micro and nanoplastics to adsorb pollutants, such as PAHs, the environment can be more negatively impacted. This results in the production of several adverse effects on the fauna and flora of the ecosystems (Gao et al., 2022). In fact, the exposure to micro and nanoplastics, together with PAHs, can cause delays in growth and in the effective absorption of gases and nutrients (Sun et al., 2021). It has been shown that the combined exposure to micro and nanoplastics in mixtures of PAHs exerts evident synergistic toxic effects on blood parameters. Additionally, the accumulation of some PAHs, such as pyrene, in some aquatic species has been reported to be enhanced by the presence of micro and nanoplastics as a consequence of their synergistic influence (Sun et al., 2020, 2021). Nonetheless, there are not sufficient evidence on this issue.

Under this scenario, this study aims to carry out an analysis of the synergistic effect of the presence of micro and nanoplastics on the toxicity associated with pyrene, and its influence on human health and the quality of aquatic ecosystems, particularly.

PAHs in the Environment

PAHs are organic pollutants that have a structure represented as a fusion of at least two stable benzene rings to which hydrogen (H) and carbon (C) atoms are attached. These compounds can be classified according to their molecular weight and aromatic rings in PAHs of low molecular weight characterized by having six rings or less; and heavy PAHs, which have more than six condensed rings in their structure (Okedere et al., 2022; Rubio-Clemente et al., 2014).

PAHs can be used in the field of electronics, functional plastics and liquid crystals (Abdel-Shafy et al., 2016). Among the general characteristics of PAHs, their high boiling and melting points stand out; which means that they are solid. They have also a low solubility in water, which decreases when the number of fused aromatic rings is increased, and they have low vapor pressure. Furthermore, PAHs exhibit high solubility in solvents, indicating highly lipophilic behavior; therefore, they are easily absorbed by the gastrointestinal tract of mammals and are rapidly distributed in tissues, tending to be located in body fat (Abdel-Shafy et al., 2016). Most PAHs are fluorescent and emit characteristic wavelengths when excited (Abdel-Shafy et al., 2016; Qazi et al., 2021). PAHs can largely absorb light in the ultraviolet (UV) region and undergo photochemical reactions, such as photodimerization and photo-oxidation (Clark et al., 2007).

These compounds are mainly generated by the incomplete combustion of organic matter. The main sources of emissions are anthropogenic, such as the processing of coal, crude oil and natural gas, aluminum and iron related-process, foundries and heating in power plants and homes (Joseph et al., 2015). Likewise, there are natural emission sources of PAHs, including open fires, the filtration of oil or coal deposits, and volcanic activity, among others (Abdel-Shafy et al., 2016). On the other hand, there is a presence of PAHs in smoked foods, and in roasted meat and fish (Bansal et al., 2015).

More than 80% of PAH emissions worldwide are attributed to developing countries, with biomass and coal burning being the activities that generate more than half of these emissions (Reizer et al., 2021). The World Health Organization (WHO) and the United States currently regulate 16 PAHs considered to be priority substances, among which pyrene is one of the most relevant (Okedere et al., 2022).

It should be noted that PAH toxicity and persistence increase according to their number of fused rings. For this reason, PAHs with high molecular weight are of particular concern. The importance given to these contaminants lies mainly in their scientifically proven mutagenic and teratogenic effects, indicating that they can react with deoxyribonucleic acid (DNA) generating mutations in different organs such as lungs, liver, breasts and skin (Reizer et al. al., 2021; Mallah et al., 2022).

Physicochemical Characteristics of Pyrene

Pyrene ($C_{16}H_{10}$) is made up of four fused aromatic rings, leading to a flat aromatic system. This compound has a high molecular weight (202.25 g/mol) and is formed from the incomplete combustion of organic chemicals (NCBI, 2022; Wang et al., 2022). As previously indicated for the rest of PAHs, pyrene is a solid compound of yellow colour (Rubio-Clemente et al., 2014). Figure 1 illustrates the chemical structure of pyrene. Furthermore, in Table 1, the main physicochemical properties of this chemical are listed (NCBI, 2022).

Pyrene has a large π-aromatic conjugate system, and most of its derivatives behave well as third-order optical nonlinear organic materials (Shi et al., 2022). It is noteworthy that pyrene has a long lifetime in the excited state (50–90 ns) and a high ease of functionalization with simple reactions such as bromination and Crafts acylation reactions (Dumur et al., 2020). Furthermore, pyrene is considered as a strong electron donor or acceptor, they can be assembled with other molecules through conjugated bridges to obtain resulting compounds that have novel characteristics and excellent properties (Shi et al., 2022).

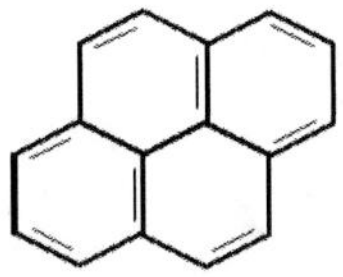

Figure 1. Chemical structure of pyrene.

Table 1. Physicochemical properties of pyrene (NCBI, 2022)

Property	Value
Molecular weight	202.25 g/mol
XLogP3	4.9
Molecular formula	$C_{16}H_{10}$
Exact mass	202.078.250.319
Monoisotopic mass	202.078.250.319
Heavy atom count	16
Boiling point	759°F at 760 mm Hg
Melting point	313°F
Flash point	>200,0°C (> 392,0°F)
Solubility in water	<1 mg/mL at 72° F
Density	1.27 g/cm³ at 73,4 °F
$LogK_{OA}$	8.8
Heat of combustion	3,878 x 10⁷ J/kg at 25°C
Heat of vaporization	3,21 x 10⁵ J/kg
Chemical safety	Environmental risk – irritating compound

Additionally, pyrene exhibits a great research interest as an organic chromophore, it is commonly used as a blue light-emitting fluorescent probe for biological analysis (Hsiao et al., 2018). Traditionally, pyrene has been studied for the design of photoluminescent materials, and its photoluminescence quantum yield has been calculated to be 0.68 when it is found in the solid state (Dumur et al., 2020). Due to its photophysical properties, pyrene is used to elaborate visible light photo-indicators (Dumur et al., 2020). Moreover, pyrene is usually known for its formation of excimers, which are products of a photochemical reaction between excited and ground-state molecules; and is often used as a probe to study micelle formation (Zhao et al., 2022). Furthermore, pyrene is utilized as a raw material in the manufacture of dyes and optical brighteners (Abdel-Shafy et al., 2016).

Emission Sources and Occurrence of Pyrene

PAHs are transported into aquatic systems through wastewater, atmospheric precipitation and surface runoff (Zhao et al., 2018), where they have a high level of toxicity for aquatic life and birds. In fact, in aquatic organisms, their metabolism can be affected, due to the photooxidation reactions associated. Moreover, these pollutants are bioaccumulative and can be biomagnified through the food chain. In this regard, they are found in living beings, including fish and shellfish, as well as in terrestrial invertebrates and vertebrates (Abdel-Shafy et al., 2016).

Concerning pyrene, it is commonly found in the environment, particularly in coastal areas and estuaries (Xie et al., 2017). Because of its low water solubility and affinity to organic material, it is easily immobilized in minerals. In addition to its high molecular weight, pyrene has unique photochemical properties, which makes it a highly active compound in redox reactions in the environment (Wang et al., 2022).

As pyrene is a high molecular weight PAH, it is easily adsorbed or immobilized on solid particles in the soil and in the aquatic environment (Wang et al., 2022), so that it is commonly found in sediments and water (Toxværd et al., 2019; Grenvald et al., 2013; Xie et al., 2017). Therefore, sediments in water represent the main sink for PAHs due to their hydrophobic nature and weak degradation (Zhao et al., 2018).

In particular, pyrene reaches the environment from natural fires, motor vehicle exhaust, cigarette smoke emissions, and smoke from stoves and ovens (Joseph et al., 2015). This compound is considered as a soot precursor (Reizer et al., 2021).

The exposure of living beings to pyrene has been studied in order to show the influence and alteration of their usual life cycle. In the research conducted by Toxværd and co-workers, the effects of the exposure of the Arctic copepod *Calanus hyperboreus* to 1–10 nM during winter were evaluated. This species of *Calanus sp.* is of vital importance for the ecosystem in the Arctic, since it supports populations of fish and mammals. In this study, egg production by this species was found to be unaffected by pyrene exposure, as *Calanus hyperboreus* has higher resistance to pyrene due to its high lipid content (Toxværd et al., 2019). On the other hand, Grenvald and co-workers conducted a study using *C. glacialis* and *C. finmarchicus*. Both species were subjected to pyrene (0.1 – 100 nM) and low temperatures. As a result, it was possible to determine a decrease in the survival of both species, since the growth of the nauplii of *C. glacialis* was reduced and the naupliar mortality of *C. finmarchicus* increased due to the exposure to this pollutant (Grenvald et al., 2013). Other authors studied bivalves (clams, mussels and oysters), commonly used as marine environmental indicators due to their longevity, infiltration feeding and resistance to extreme conditions (Xie et al., 2017). It is noteworthy that these species are bioaccumulative and have low tolerance to PAHs, so that when they are exposed to pyrene, they presented significant oxidative damage and a reduction in the total number of hemocytes and in the stability of the cell membrane, as well as in the phagocytic activity. Moreover, an increase in lipid peroxidation was also observed (Xie et al., 2017).

Micro and Nanoplastics in the Environment

The environmental problem with respect to plastics is a worldwide issue, representing one of the highest priority challenges nowadays (Gao et al., 2022). Plastic is widely used in industry, agriculture, medicine and daily life due to its characteristics of durability and low weight and production cost, among others (Patil et al., 2022). Plastics are classified according to their size into macroplastics (5–50 cm), mesoplastics (0.5–5 cm), microplastics (1 μm < size < 5 mm) and nanoplastics (< 1000 nm) (Priya et al., 2022; Monikh et al., 2022). The incomplete biodegradability of plastic polymers results in their decomposition into smaller particles, increasing their persistence in the environment (Patil et al., 2022).

Micro and nanoplastics are ubiquitous, since they are present in all ecosystems and habitats. Micro and nanoplastics are not only generated by the fragmentation of larger plastics, but they are also manufactured directly to be used in paints, toothpastes, cosmetics and cleaners, among other items (Banerjee et al., 2021). In the literature, the presence of these materials is reported in the ocean, freshwater, domestic wastewater, river and lake sediments, soil, and atmosphere (Zhang et al., 2022). In fact, in 2016, pollution with micro and nanoplastics is included as the second most important scientific topic in environmental and ecological sciences in the second assembly of the United Nations (Zhang et al., 2022). Given the importance recently given to plastic pollution and its influence on the environment, strategies have been developed for the environmental management of this waste, such as recycling and incineration for energy recovery, with the main objective of promoting a proper management of this waste at the end of their useful life, which ranges between 1 and 50 years depending on the product (Caldwell et al., 2022).

Physicochemical Properties of Micro and Nanoplastics

The concept of micro-nanoplastics was coined for the first time in 2004 by Thompson, et al., indicating that they are organic objects with a size < 5 mm (Thompson et al., 2004).

These substances are recognized as ubiquitous anthropogenic pollutants (Parolini et al., 2021). Due to their minute size, micro-nanoplastics cannot be removed from the environment, as their detection and quantification is practically impossible (Meng et al., 2020). Attempts to separate particles from environmental matrices often result in incomplete separations or damage to the particles (Caldwell et al., 2022), making them even smaller and thus increasing their hazard and associated risk. In the environment, micro-nanoplastics can easily alter their composition and physical properties, modifying the structure of the polymers and their additives, generating alterations in the structures of their chemical bonds due to abiotic and biotic factors (Parolini et al., 2021).

Micro and nanoplastics represent an alarm for the health of ecosystems and humankind, due to their high surface area and tiny size, so that they can adsorb other pollutants of concern such as antibiotics, heavy metals and PAHs (Zhang et al., 2022), among which pyrene is of special interest. Micro and nanoplastics are lightweight, made of strong and corrosion-resistant materials, and have low thermal and electrical conductivity (Meng et al., 2020). Depending on the behavior of the polymers used to make them during the initial exposure to heat, plastics can be classified as thermosetting or thermoplastic products. Thermoset plastics, are those plastics that when heated and undergo chemical changes, resulting in the formation of a three-dimensional crosslinked matrix, which cannot be melted or reformed. In turn, thermoplastics are those plastics that when heated above a specific temperature become pasty, but below that temperature they are solid; thanks to this property they can be melted and reformed a large number of times (Caldwell et al., 2022).

Micro and nanoplastics can be also classified according to their source in primary and secondary: i) the former type refers to those plastic particles that are in their state without any alteration since their production; and ii) the later refers to those plastics derived from the primary ones as a result of wear and tear or the effects of environmental degradation due to physicochemical and biological reactions, and abiotic weathering (Kershaw et al., 2019). The degradation of micro and nanoplastic particles mainly depends on the type of polymer used in its production, environmental factors such as UV radiation and the exposure to high temperatures combined with the oxygen of the atmosphere cause thermo- and photo-oxidation. On the other hand, exposure to abrasion, wind and waves are agents with degradation capacity also widely discussed in the scientific literature (Caldwell et al., 2022).

Production and Occurrence of Micro and Nanoplastics

The main sources of micro and nanoplastics are larger plastic items, including bottles, caps, bags and wrappers (Parolini et al., 2021), that are fragmented through physical, chemical and biological reactions. It is noteworthy that approximately 50% of the plastic is single-use, which is discarded without a proper management, causing an increase in the environmental pollution, estimating 12 billion metric tons of plastic waste by 2050 (Patil et al., 2022).

Among the sources of micro and nanoplastics in fresh water, synthetic textiles, personal care and hygiene products, raw materials in industry and inadequate treatment and disposal of waste stand out (Li et al., 2020). Once used, plastic waste reaches the marine environment by rivers, drainage systems and/or the wind.

Plastic is considered as an omniscient pollutant in the entire marine environment, from the coasts to the depths of the ocean. Indeed, it has been reported that more than half of marine mammal species and all known species of sea turtles have ingested micro and nanoparticles of plastics (Gall et al., 2015). On the other hand, the main source of pollution of micro and nanoplastics in the soil is the residual sludge from wastewater treatment plants (WWTPs), highlighting the synthetic fibers that cannot be removed and reach the environment, as well as the non-existent management of some of these materials due to the mishandling provided by society (Chae et al., 2018). In the atmosphere, micro-nanoplastics enter mainly through combustion gas emissions, road traffic and the friction of tires with the pavement (Patil et al., 2022). In the hydraulic transport of WWTPs, these types of particles are also released into the atmosphere. Due to their low density, they are very likely to be spread in the air and deposited in water systems, so these particles can easily come into direct contact and ingestion with humans and other living beings, causing health effects (Zhang et al., 2022).

The micro-nanoplastics present in the environment are prone to adsorbing heavy metals and all kinds of organic contaminants, particularly those that are hydrophobic in nature, including pyrene. This leads to a greater damage within the ecosystem and its organisms (Zhang et al., 2022; Zeb et al., 2022).

In all the environments studied, the problem of micro-nanoplastics is considered a common factor, since due to their small size, they are easily incorporated by different animal and vegetal species, causing negative repercussions on flora and fauna health (Zhang et al., 2022; Gall et al., 2015). Plastics in the environment represent an inconvenience for the development of animal life, since birds can get entangled in this type of materials, and marine species such as turtles ingest them easily, which is lethal for the species and diversity of the ecosystem (Meng et al., 2020). Not only animals are affected by the presence of these emerging pollutants, micro-nanoplastics can also alter the growth of microalgae and thus affect the entire aquatic ecosystem. In addition, microalgae through the biofouling of micro-nanoplastics can increase their density and modify their characteristics (Mateos-Cárdenas et al., 2021). Furthermore, in the environment, the presence of micro and nanoplastics can generate an accumulation in microbial populations, leading to the creation of biofilms (Zhang et al., 2022), and increasing the risk of pollution by pathogen load.

On the other hand, micro-nanoplastics have the capacity to adsorb a large amount of organic pollutants, as mentioned above, increasing their bioavailability, toxicity and dispersion in the environment (Okoye et al., 2022). Micro and nanoplastics, due to their smaller size, have a larger surface area and a higher load capacity to adsorb both organic and inorganic pollutants (Liu et al., 2022), among which PAHs are found. Once PAHs are distributed in the environment, they can be adsorbed by micro and nanoplastics due to the large presence of these ECs in the different environmental matrices. This sorption depends on the properties of both components, being the hydrophobicity and molecular polarity of PAHs the characteristics that govern the adsorption process between PAHs, such as pyrene, and micro and nanoplastics. On the other hand, concerning to the properties of micro and nanoplastics favouring the adsorption of PAHs, their type, size and crystallinity are highlighted (Yang et al., 2021). Particularly, pyrene has been reported to have an especial affinity with some types of micro-nanoplastics such as polyethylene, polystyrene and polyvinyl chloride, since pyrene hydrophobicity has been directly correlated with its adsorption coefficients (Yang et al., 2021).

Synergetic Adverse Effects Associated with the Presence of Micro and Nanoplastics

The main routes of exposure to micro-nanoplastics are inhalation, dermal absorption and ingestion, giving rise to conditions in humans, such as lung damage, tissue lodging of the particles in the spleen and liver, as well as liver and gastrointestinal damage (Zarus et al., 2021).

When humans are exposed to plastic dust, it tends to cause difficulty in breathing, coughing, increased expectoration, irritation of the respiratory tract and reduced lung capacity (Banerjee et al., 2021). When plastic fibers reach lung tissues, they can cause lung inflammation and genotoxicity (Zarus et al., 2021). Gasperi et al. determined the alterations suffered by the exposure to micro and nanoplastics, especially fibrous particles. The fibers can be natural or artificial, and the adverse effects that these fibers can generate when entering our respiratory system depend mainly on their size. It is important to emphasize that the fibrous particles that are deposited in the lung are classified as respirable, while those that are deposited in the upper respiratory tract are called inhalable. The aforementioned authors found that the fibrous micro-nanoplastic particles are persistent in the body and are difficult to expel. Regardless of the dose, any exposure to these pollutants has been reported to generate lung inflammation after chronic inhalation, leading to fibrosis when longer fibers enter the living being (Gasperi et al., 2018). In turn, Banerjee and co-workers found that polyethylene and polycarbonate fibers can persist in the pulmonary system for up to 6 months (Banerjee et al., 2021), and that the liver was one of the most affected organs when exposed to this plastic materials. The authors investigated the effects of polystyrene micro-nanoparticles in a size range of 50–5000 nm and levels of 0.1–100 μg/mL on HepG2 liver cells for an exposure time from 1 to 24 h. Smaller particles were found to be more toxic to hepatocytes, while larger ones induced apoptotic cell death or an inflammatory response, depending on the exposure time. Thus, the toxicity and damage caused by polystyrene particles in liver cells were concluded to be dependent on their size, the concentration to which the organ was exposed and the exposure times (Banerjee et al., 2022).

On the other hand, micro-nanoplastics, as indicated above, can interact with other types of organic pollutants, being potential chemical carriers and altering their toxicity. Micro-nanoplastics are determining factors in the absorption and transport of hydrophobic contaminants. Polymers, including polystyrene, polyethylene and polypropylene, have been informed to exhibit a high capacity to absorb PAHs (Avio et al., 2015), such as pyrene. Avio and co-workers evaluated the effect of two polymers (polyethylene and polystyrene) impregnated with pyrene at different concentrations (0.5, 5 and 50 μg/L), being the two former levels the closest to those usually found in the environment for pyrene. The later concentration could be associated with an extreme event, like an oil spill. In this ecotoxicological research, mussels (*Mytilus galloprovincialis*) were used. The results obtained revealed that pyrene molecules present in the micro-nanoplastics were transferred to the organisms, being bioaccumulated in the tissues. This study demonstrated the great capacity of the micro and nanoplastics to adsorb and transport organic pollutants in the aqueous medium and their influence on the transferring and bioaccumulation in living organisms (Avio et al., 2015).

Polystyrene is one of the main polymers used in ecotoxicological experimentation with pyrene. Recently, Liu and co-workers determined the toxicity of pyrene combined with micro-nanoplastics on earthworms (*Eisenia fetida*). For this study, 10-μm microplastics, 100-nm nanoplastics and pyrene-contaminated soil at concentrations of 10 and 100 mg/kg were used by exposing earthworms. Pyrene itself affects the growth rate, antioxidant systems, enzyme activity, reproductive system and DNA of the earthworms. When pyrene was combined with micro-nanoplastics, exfoliation and deformation of the intestinal tissue was observed, thinning the intestinal wall, and increasing the intestinal epithelial cells and spaces. In turn, micro-nanoplastics were found to increase pyrene accumulation in earthworms by damaging intestinal tissue and disrupting detoxification functions. Therefore, the co-exposure of mussels to both contaminants under the experimental conditions studied here increased toxicity due to the combination of adverse effects (Liu et al., 2022).

Contrarily, the presence of micro and nanoparticles of plastic can reduce the bioavailability of PAHs to affect living beings negatively. In the research carried out by Barbosa and co-workers, it was evidenced that the exposure of embryos and larvae of the zebrafish (*Danio rerio*) to a mixture made up of 36 PAHs showed cardiac malformations, jaw, fin and tail deformities. In turn, the vascular development of the brain was also altered due to the mixture of pollutants. However, when polystyrene nanoparticles were mixed, it was found that the larvae did not exhibit developmental defects. Additionally, it was observed that when the larvae were exposed to the mixture of PAHs and nanoparticles simultaneously, the occurrence of developmental deformations in fishes decreased. This fact could be ascribed to the decrease in PAH concentrations given the surface adsorption by the polystyrene particles of these pollutants; i.e., the toxicity of the PAH mixture was reduced (Barbosa et al., 2020).

In a study conducted by Trevisan et al., zebrafish embryos and larvae were also exposed to a mixture of PAHs ranging from 5.07 to 25.36 μg/L and/or polystyrene nanoparticles with a concentration of 0.1–10 mg/L. Similarly, no effects were observed when nanoparticles were individually used; while when zebrafish larvae were exposed to the mixture of PAHs, they exhibit jaw deformities, tail and fin malformation and heart development problems.

During the co-exposure experiments, it was found that the nanoparticles decreased the negative effects associated with PAH exposure, and this phenomenon was ascribed to the adsorption of PAHs to the nanoparticle surface and the agglomeration of nanoparticles, being more difficult to be incorporated into the organisms and reducing the amount of PAHs to be in contact with the living beings (Trevisan et al., 2019).

In this regard, and in order to discern the capacity of micro and nanoplastics to increase the potential adverse effect ascribed to PAHs, such as pyrene, further studies are needed.

Conclusion

The pollution problem related to the presence of micro-nanoplastics and PAHs, such as pyrene, in all environmental matrices, especially in water, represents an alarm in terms of environmental quality and human and living being health. Micro and nanoplastics carry out functions of collection and transport of organic pollutants, including pyrene, increasing the health risks related to their carcinogenic, mutagenic and teratogenic potential. Additionally, because of the properties of micro and nanoplastics, massive effects have been reported on the respiratory and cardiovascular systems of human beings, and on the optimal development of marine and terrestrial species due to co-exposure to these pollutants; although, the opposite effect has been also informed. Therefore, further research and studies about the risks involved in being in contact with micro and nanoplastic particles polluted with PAHs, like pyrene, are required because there is scarce information and it is a field of study of growing concern broadly.

Acknowledgment

Authors acknowledge the financial support provided by Universidad de Antioquia (Estrategia de Sostenibilidad 2020-2021. ES84190067).

References

Abdel-Shafy, H. I., and Mansour, M. S. (2016). A review on polycyclic aromatic hydrocarbons: source, environmental impact, effect on human health and remediation. *Egyptian Journal of Petroleum*, 25(1), 107-123.

Anbumani, S., and Kakkar, P. (2018). Ecotoxicological effects of microplastics on biota: a review. *Environmental Science and Pollution Research*, 25(15), 14373-14396.

Avio, C. G., Gorbi, S., Milan, M., Benedetti, M., Fattorini, D., d'Errico, G., Pauletto, M., Bargellini, L., and Regoli, F. (2015). Pollutants bioavailability and toxicological risk from microplastics to marine mussels. *Environmental Pollution*, 198, 211-222.

Aznar-Sánchez, J. A., Velasco-Muñoz, J. F., Belmonte-Ureña, L. J., and Manzano-Agugliaro, F. (2019). The worldwide research trends on water ecosystem services. *Ecological Indicators*, 99, 310-323.

Banerjee, A., and Shelver, W. L. (2021). Micro-and nanoplastic induced cellular toxicity in mammals: A review. *Science of The Total Environment*, 755, 142518.

Banerjee, A., Billey, L. O., McGarvey, A. M., and Shelver, W. L. (2022). Effects of polystyrene micro/nanoplastics on liver cells based on particle size, surface functionalization, concentration and exposure period. *Science of The Total Environment*, 155621.

Bansal, V., and Kim, K. H. (2015). Review of PAH contamination in food products and their health hazards. *Environment International*, 84, 26-38.

Barbosa, F., Adeyemi, J. A., Bocato, M. Z., Comas, A., and Campiglia, A. (2020). A critical viewpoint on current issues, limitations, and future research needs on micro-and nanoplastic studies: From the detection to the toxicological assessment. *Environmental Research*, 182, 109089.

Boots, B., Russell, C. W., and Green, D. S. (2019). Effects of microplastics in soil ecosystems: above and below ground. *Environmental Science & Technology*, 53(19), 11496-11506.

Caldwell, J., Blanco, P. T., Lehner, R., Lubskyy, A., Ortuso, R. D., Rothen-Rutishauser, B., and Petri-Fink, A. (2022). The micro-, submicron-, and nanoplastic hunt: A review of detection methods for plastic particles. *Chemosphere*, 133514.

Chae, Y., and An, Y. J. (2018). Current research trends on plastic pollution and ecological impacts on the soil ecosystem: A review. *Environmental Pollution*, 240, 387-395.

Clark, C. D., De Bruyn, W. J., Ting, J., and Scholle, W. (2007). Solution medium effects on the photochemical degradation of pyrene in water. *Journal of Photochemistry and Photobiology A: Chemistry*, 186(2-3), 342-348.

Clemente, P., Calvache, M., Antunes, P., Santos, R., Cerdeira, J. O., & Martins, M. J. (2019). Combining social media photographs and species distribution models to map cultural ecosystem services: The case of a Natural Park in Portugal. *Ecological Indicators*, 96, 59-68.

Dumur, F. (2020). Recent advances on pyrene-based photoinitiators of polymerization. *European Polymer Journal*, 126, 109564.

Feng, T., Wang, C., Hou, J., Wang, P., Liu, Y., Dai, Q., Yangyang, Y., and You, G. (2018). Effect of inter-basin water transfer on water quality in an urban lake: A combined water quality index algorithm and biophysical modelling approach. *Ecological Indicators*, 92, 61-71.

Gall, S. C., and Thompson, R. C. (2015). The impact of debris on marine life. *Marine Pollution Bulletin*, 92(1-2), 170-179.

Gao, D., Liu, X., Junaid, M., Liao, H., Chen, G., Wu, Y., and Wang, J. (2022). Toxicological impacts of micro(nano)plastics in the benthic environment. *Science of The Total Environment*, 155620.

Gasperi, J., Wright, S. L., Dris, R., Collard, F., Mandin, C., Guerrouache, M., Langlois V., Kelly, F. J., and Tassin, B. (2018). Microplastics in air: are we breathing it in? *Current Opinion in Environmental Science & Health*, 1, 1-5.

Grenvald, J. C., Nielsen, T. G., and Hjorth, M. (2013). Effects of pyrene exposure and temperature on early development of two co-existing Arctic copepods. *Ecotoxicology*, 22(1), 184-198.

Hsiao, S. H., and Huang, Y. P. (2018). Redox-active and fluorescent pyrene-based triarylamine dyes and their derived electrochromic polymers. *Dyes and Pigments*, 158, 368-381.

Huang, W., Song, B., Liang, J., Niu, Q., Zeng, G., Shen, M., Deng, J., Luo, Y., Wen, X. and Zhang, Y. (2021). Microplastics and associated contaminants in the aquatic environment: A review on their ecotoxicological effects, trophic transfer, and potential impacts to human health. *Journal of Hazardous Materials*, 405, 124187.

Joseph P. Sullivan, Chapter 32 - Wildlife Toxicity Assessment for Pyrene, Editor(s): Marc A. Williams, Gunda Reddy, Michael J. Quinn, Mark S. Johnson, *Wildlife Toxicity Assessments for Chemicals of Military Concern*, Elsevier, 2015, Pages 599-616, ISBN 9780128000205,

Kershaw, P., Turra, A., and Galgani, F. (2019). Guidelines for the Monitoring and Assessment of Plastic Litter in the Ocean. *GESAMP Reports and Studies*. Retrieved 4 June, 2022 from: https://wesr.unep.org/media/docs/marine_plastics/une_science_dvision_gesamp_reports.pdf.

Li, C., Busquets, R., and Campos, L. C. (2020). Assessment of microplastics in freshwater systems: A review. *Science of the Total Environment*, 707, 135578.

Liu, Y., Xu, G., and Yu, Y. (2022). Effects of polystyrene microplastics on accumulation of pyrene by earthworms. *Chemosphere*, 296, 134059.

López, A. F., Fabiani, M., Lassalle, V. L., Spetter, C. V., and Severini, M. F. (2022). Critical review of the characteristics, interactions, and toxicity of micro/nanomaterials pollutants in aquatic environments. *Marine Pollution Bulletin*, 174, 113276.

Mallah, M. A., Changxing, L., Mallah, M. A., Noreen, S., Liu, Y., Saeed, M., Xi, H., Ahmed, B., Feng, F., Mirjat, A. A., Wang W., Jabar, A., Naveed, M., Li, J., and Zhang, Q. (2022). Polycyclic aromatic hydrocarbon and its effects on human health: an updated review. *Chemosphere*, 133948.

Mateos-Cárdenas, A., van Pelt, F. N., O'Halloran, J., and Jansen, M. A. (2021). Adsorption, uptake and toxicity of micro-and nanoplastics: Effects on terrestrial plants and aquatic macrophytes. *Environmental Pollution*, 284, 117183.

Meng, Y., Kelly, F. J., and Wright, S. L. (2020). Advances and challenges of microplastic pollution in freshwater ecosystems: A UK perspective. *Environmental Pollution*, 256, 113445.

Monikh, F. A., Durão, M., Kipriianov, P. V., Huuskonen, H., Kekäläinen, J., Uusi-Heikkilä, S., Uurasjärvi, J., Akkanen, J., and Kortet, R. (2022). Chemical composition and particle size influence the toxicity of nanoscale plastic debris and their co-occurring benzo (α) pyrene in the model aquatic organisms *Daphnia magna* and *Danio rerio*. *NanoImpact*, 100382.

National Center for Biotechnology Information (2022). *PubChem Compound Summary for CID 31423, Pyrene*. Retrieved June 4, 2022 from https://pubchem.ncbi.nlm.nih.gov/compound/Pyrene.

Okedere, O. B., and Elehinafe, F. B. (2022). Occurrence of Polycyclic Aromatic Hydrocarbons in Nigeria's Environment: A Review. *Scientific African*, e01144.

Okoye, C. O., Addey, C. I., Oderinde, O., Okoro, J. O., Uwamungu, J. Y., Ikechukwu, C. K., Ejeromedoghene, O., and Odii, E. C. (2022). Toxic Chemicals and Persistent Organic Pollutants Associated with Micro-and Nanoplastics Pollution. *Chemical Engineering Journal Advances*, 100310.

Paredes, I., Ramírez, F., Forero, M. G., and Green, A. J. (2019). Stable isotopes in helophytes reflect anthropogenic nitrogen pollution in entry streams at the Doñana World Heritage Site. *Ecological Indicators*, 97, 130-140.

Parolini, M., Ortenzi, M. A., Morelli, C., and Gianotti, V. (2021). Emerging use of thermal analysis in the assessment of micro (nano) plastics exposure. *Current Opinion in Toxicology*, 28, 38-42.

Patil, S. M., Rane, N. R., Bankole, P. O., Krishnaiah, P., Ahn, Y., Park, Y. K., Yadav, K. K., Amin, M. A., and Jeon, B. H. (2022). An assessment of micro-and nanoplastics in the biosphere: A review of detection, monitoring, and remediation technology. *Chemical Engineering Journal*, 430, 132913.

PlasticsEurope (2019). *Plastics – The Facts 2019*. An Analysis of European Latest Plastics Production, Demand and Waste Data Association of Plastic Manufacturers. Retrieved 4 June, 2022 from: https://plasticseurope.org/wp-content/uploads/2021/10/2019-Plastics-the-facts.pdf.

Priya, A. K., Jalil, A. A., Dutta, K., Rajendran, S., Vasseghian, Y., Qin, J., and Soto-Moscoso, M. (2022). Microplastics in the environment: Recent developments in characteristic, occurrence, identification and ecological risk. *Chemosphere*, 134161.

Qazi, F., Shahsavari, E., Prawer, S., Ball, A. S., and Tomljenovic-Hanic, S. (2021). Detection and identification of polyaromatic hydrocarbons (PAHs) contamination in soil using intrinsic fluorescence. *Environmental Pollution*, 272, 116010.

Reizer, E., Viskolcz, B., and Fiser, B. (2021). Formation and growth mechanisms of polycyclic aromatic hydrocarbons: A mini-review. *Chemosphere*, 132793.

Rubio-Clemente, A., Torres-Palma, R. A., and Peñuela, G. A. (2014). Removal of polycyclic aromatic hydrocarbons in aqueous environment by chemical treatments: a review. *Science of The Total Environment*, 478, 201-225.

Salehi, M. (2022). Global water shortage and potable water safety; Today's concern and tomorrow's crisis. *Environment International*, 158, 106936.

Shi, Y., Jia, J., Sun, J., Yin, A., Zhao, M., and Song, Y. (2022). Study of ultrafast nonlinear optical response and transient dynamics of pyrene derivatives with intramolecular charge transfer characteristics. *Optical Materials*, 128, 112378.

Sun, B., Li, Q., Zheng, M., Su, G., Lin, S., Wu, M., Li, C., Wang, Q., Tao, Y., Dai, L., Qin, Y., and Meng, B. (2020). Recent advances in the removal of persistent organic pollutants (POPs) using multifunctional materials: a review. *Environmental Pollution*, 265, 114908.

Sun, S., Shi, W., Tang, Y., Han, Y., Du, X., Zhou, W., Zhang, W., Sun, C., and Liu, G. (2021). The toxic impacts of microplastics (MPs) and polycyclic aromatic hydrocarbons (PAHs) on haematic parameters in a marine bivalve species and their potential mechanisms of action. *Science of The Total Environment*, 783, 147003.

Thompson, R. C., Olsen, Y., Mitchell, R. P., Davis, A., Rowland, S. J., John, A. W., McGonigle, D., and Russell, A. E. (2004). Lost at sea: where is all the plastic? *Science*, 304(5672), 838-838.

Toxværd, K., Dinh, K. V., Henriksen, O., Hjorth, M., and Nielsen, T. G. (2019). Delayed effects of pyrene exposure during overwintering on the Arctic copepod *Calanus hyperboreus*. *Aquatic Toxicology*, 217, 105332.

Trevisan, R., Voy, C., Chen, S., and Di Giulio, R. T. (2019). Nanoplastics decrease the toxicity of a complex PAH mixture but impair mitochondrial energy production in developing zebrafish. *Environmental Science & Technology*, 53(14), 8405-8415.

United Nations World Water Assessment Programme (WWAP) (2015). Water for a sustainable world. *The United Nations World Water Development Report*. Paris, UNESCO.

Wang, C., Hao, Z., Huang, C., Wang, Q., Yan, Z., Bai, L., Ji, H., and Li, D. (2022). Drinking water treatment residue recycled to synchronously control the pollution of polycyclic aromatic hydrocarbons and phosphorus in sediment from aquatic ecosystems. *Journal of Hazardous Materials*, 431, 128533.

Wang, L., Li, C., Jiao, B., Li, Q., Su, H., Wang, J., and Jin, F. (2018). Halogenated and parent polycyclic aromatic hydrocarbons in vegetables: levels, dietary intakes, and health risk assessments. *Science of The Total Environment*, 616, 288-295.

Wang, X., Teng, Y., Wang, X., Li, X., and Luo, Y. (2022). Microbial diversity drives pyrene dissipation in soil. *Science of The Total Environment*, 153082.

Wang, Z., Wang, F., Xiang, L., Bian, Y., Zhao, Z., Gao, Z., Chen, J., Shaeffer, A., Jiang, X., and Dionysiou, D. D. (2022). Degradation of mineral-immobilized pyrene by ferrate oxidation: Role of mineral type and intermediate oxidative iron species. *Water Research*, 217, 118377.

Xie, J., Zhao, C., Han, Q., Zhou, H., Li, Q., and Diao, X. (2017). Effects of pyrene exposure on immune response and oxidative stress in the pearl oyster, *Pinctada martensii*. *Fish & Shellfish Immunology*, 63, 237-244.

Yan, J., Wang, L., Fu, P. P., and Yu, H. (2004). Photomutagenicity of 16 polycyclic aromatic hydrocarbons from the US EPA priority pollutant list. *Mutation Research/Genetic Toxicology and Environmental Mutagenesis*, 557(1), 99-108.

Yang, C., Wu, W., Zhou, X., Hao, Q., Li, T., and Liu, Y. (2021). Comparing the sorption of pyrene and its derivatives onto polystyrene microplastics: Insights from experimental and computational studies. *Marine Pollution Bulletin*, 173, 113086.

Zarus, G. M., Muianga, C., Hunter, C. M., and Pappas, R. S. (2021). A review of data for quantifying human exposures to micro and nanoplastics and potential health risks. *Science of The Total Environment*, 756, 144010.

Zeb, A., Liu, W., Shi, R., Lian, Y., Wang, Q., Tang, J., and Lin, D. (2022). Evaluating the knowledge structure of micro-and nanoplastics in terrestrial environment through scientometric assessment. *Applied Soil Ecology*, 177, 104507.

Zhang, Y., Li, Y., Su, F., Peng, L., and Liu, D. (2022). The life cycle of micro-nano plastics in domestic sewage. *Science of The Total Environment*, 802, 149658.

Zhang, Y., Zhang, L., Huang, Z., Li, Y., Li, J., Wu, N., He, J., Zhang, Z., Liu, Y., and Niu, Z. (2019). Pollution of polycyclic aromatic hydrocarbons (PAHs) in drinking water of China: Composition, distribution and influencing factors. *Ecotoxicology and Environmental Safety*, 177, 108-116.

Zhao, Y., Xu, P., Zhang, K., Schönherr, H., and Song, B. (2022). Strong emission of excimers realized by dense packing of pyrenes in tailored bola-amphiphile nano assemblies. *Cell Reports Physical Science*, 100734.

Zhao, Z., Qin, Z., Zhang, D., and Hussain, J. (2018). Dissipation characteristics of pyrene and ecological contribution of submerged macrophytes and their biofilms-leaves in constructed wetland. *Bioresource Technology*, 267, 158-166.

Chapter 4

Self-Assembling Supramolecular Structures of Pyrene

Sandeep Kumar[*]

Department of Chemistry, Nitte Meenakshi Institute of Technology,
Yelahanka, Bangalore, India;
Raman Research Institute, Bangalore, India

Abstract

Self-assembling supramolecular structures formed by functionalized disc-shaped molecules are commonly known as discotic liquid crystals (DLCs). Upon appropriate substitution, pyrene derivatives exhibit liquid crystalline properties. Pyrene as a core for DLCs was discovered in 1995. During the past three decades, a number for liquid crystalline derivatives of pyrene have been synthesized and characterized. In this chapter, the synthesis, characterization and physical properties of some self-assembling supramolecular structures of pyrene core are presented.

Keywords: pyrene, supramolecular, discotic, liquid crystals

Introduction

Disc-shaped molecules like benzene, triphenylene, pyrene, dibenzopyrene, coronene, etc., upon appropriate substitution, self-assemble in to various supramolecular structures which are commonly known as discotic liquid crystals (DLCs).

---

[*] Corresponding Author's Email: skumar.sandeep@gmail.com.

In: Pyrene: Chemistry, Properties and Uses
Editor: Charles R. Howe
ISBN: 979-8-88697-670-0
© 2023 Nova Science Publishers, Inc.

These materials are of significant interest as they self-organize into highly anisotropic and ordered supramolecular architectures using non-covalent interactions such as, π–π interactions, van der Waals interactions, ionic, dipolar and quadrupolar interactions, etc. Such systems possess both order and mobility; order of crystalline solids while mobility of isotropic liquids [1]. DLCs primarily form two types of mesophases; nematic having only orientational order but no positional order and columnar having both orientational and positional order (Figure 1). The columnar phases are very common followed by the nematic phases. Other phases like cubic phase, smectic phase etc. have also been seen in DLCs but rarely. The molecules forming DLCs are generally possess two parts; one polar rigid core which is surrounded by flexible aliphatic chains connected directly to the core or via some linking atoms or groups. The mesomorphism occurs due to microphase segregation of two non-compatible polar and non-polar parts. Due to strong π-π interactions of rigid core, these molecules spontaneously self-organize into columns and form various structures like hexagonal, rectangular, tetragonal, oblique, etc. Being liquid-like in nature, these columnar structures can be oriented parallel (planer alignment) or perpendicular (homeotropic alignment) to the surface. When these columnar phases are placed in between two electrodes and charges are created chemically or photochemically, the electrons or holes migrate from one electrode to another efficiently via hoping through the aligned molecules.

The charge mobility along the columns was observed to be several orders of magnitude higher than in the other direction and, therefore, they are called one-dimensional conductors [2]. Because of this property, the use of DLCs in semiconductor devices like photovoltaic solar cells, light emitting diodes, photoconductors, sensors, thin-film transistors, etc., has been extensively studied, e.g., [3-26].

Pyrene (Figure 2) is a member of polycyclic aromatic hydrocarbon (PAH) family comprising of four fused benzene rings. It is a colourless or light yellow solid obtained from heating of organic molecules [27, 28]. It can be obtained from combustion of acetylene and hydrogen or destructive distillation of coal tar [29].

Its synthesis from o,o'-ditolyl appeared in the literature in 1913 [30]. Subsequently, because of its industrial applications, several synthetic routes have been developed [31].

Pyrene received tremendous interest because of its high fluorescence quantum yield and its various derivatives have been extensively studied as fluorophores in chemical biology [32-34].

Many pyrene derivatives have also been used in various devices like photoconductors, sensors, organic light emitting diodes, fluorescent polymers, genetic probes, etc. [35-45].

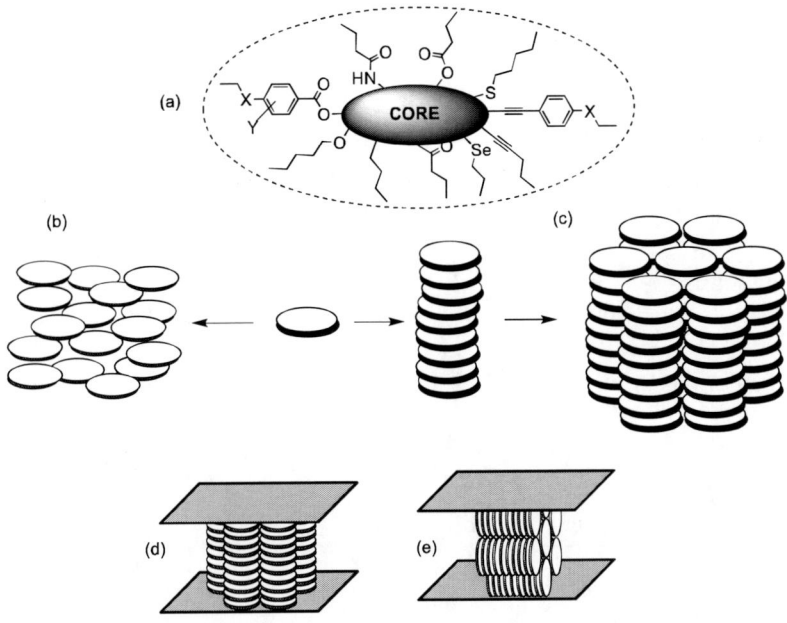

Figure 1. (a) General template of discotic liquid crystal; (b) structure of discotic nematic phase; (c) self-assembly of discs into columns and structure of hexagonal columnar phase; (d) homeotropic alignment of DLCs; (e) planar alignment of DLCs.

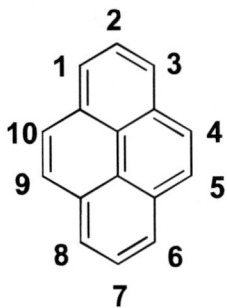

Figure 2. Chemical structure of pyrene molecule.

Pyrene Liquid Crystals:
Synthesis, Mesomorphism and Physical Properties

The incorporation of pyrene molecule in liquid crystals is dated back to 1969 when Tomkiewicz and Weinreb studied "on the decay time of pyrene in a liquid crystal" [46]. The very first discovered liquid crystal i.e., cholesterol benzoate was used and the decay time of pyrene was found to be strongly dependent on temperature. Upon appropriate substitution, pyrene derivatives exhibit liquid crystalline properties. Pyrene molecule (Figure 1) can be conveniently substituted at 1,3,6 and 8-positions to generate discotic liquid crystals. Bromination or chlorination of pyrene 1 yields 1,3,6,8-tetrahalopyrene 2 in excellent yield (Figure 3) [47, 48]. Compound 2 on treatment with fuming sulphuric acid followed by hydrolysis gives 3,6-dihydroxypyrene-1,6-quinone 3. Reductive esterification of 3 using different acid chlorides provides various 1,3,6,8-tetraalkanoyloxypyrene derivatives 4.

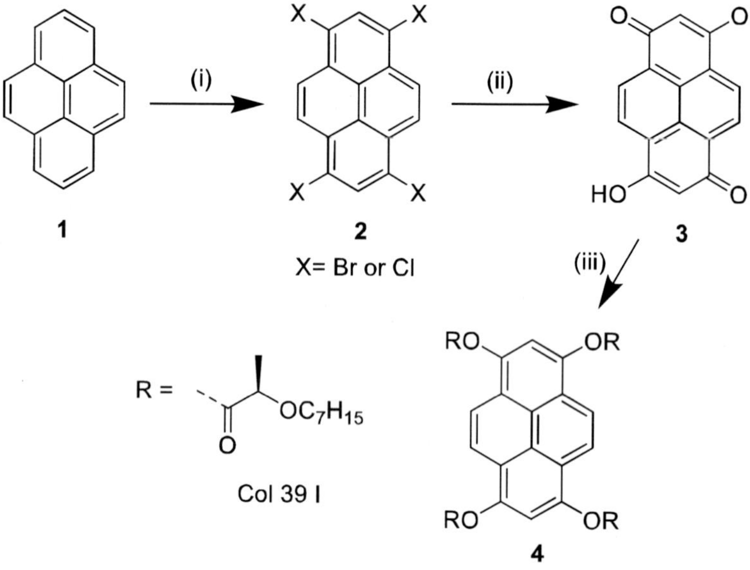

Figure 3. Synthesis of pyrene tetraesters; (i) Br2, nitrobenzene or Cl2, C2H2Cl4; (ii) 25% H2SO4-SO3; 40% H2SO4, H2O; (iii) RCOCl, Zn, DMAP, THF, pyridine.

This way, a number of pyrene tetraesters with long linear alkyl chains were prepared and characterized. However, only one derivative 4 with long branched chains was found to be liquid crystalline in virgin state. Its clearing temperature was reported to be 39°C and on slow cooling it exhibits a monotropic columnar phase below 34°C. This material displays slow ferroelectric switching. Other lower homologues with normal or branched chains periphery were found to be non-mesomorphic; however, columnar phases can be induced via charge-transfer complexation with TNF.

On the other hand, short chain alkyl ester of pyrene 1,3,6,8-tetracarboxylic acid (Figure 4), 5a, 5b and racemic 2-ethylhexyl ester, 5c, were found to be liquid crystalline [49-51]. Compound 5a exhibits a monotropic columnar phase while its higher homologue 5b display an enantiotropic columnar phase. The use of branch alkyl chain reduces the isotropic temperature significantly and the compound 5c (pyrene-1,3,6,8-tetracarboxylic tetra(2-ethylhexyl)ester) clears at about 94°C. This compound was also reported to possess a plastic phase at lower temperature [52].

The molecular dynamics of 5c is studied by dielectric relaxation and specific heat spectroscopy. Dielectric spectroscopy shows 3 processes: a β-relaxation at low temperatures and an α-relaxation in the temperature range of the mesophase followed by conductivity. The dielectric α-relaxation is assigned to a restricted glassy dynamics in the plastic crystal as well as in the liquid crystalline phase [52]. An organic light emitting diode (OLED) device has been fabricated using these materials [50].

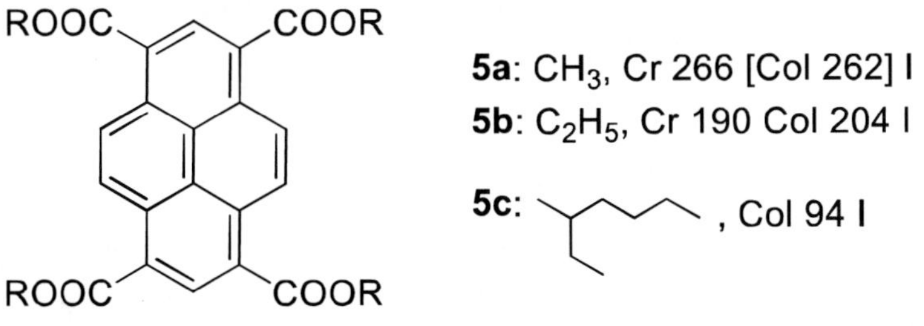

5a: CH_3, Cr 266 [Col 262] I
5b: C_2H_5, Cr 190 Col 204 I
5c: , Col 94 I

Figure 4. Liquid crystalline alkyl ester of pyrene 1,3,6,8-tetracarboxylic acid.

Several other tetrasubstituted pyrene derivatives such as, ethers 6, thioethers 7, esters and benzoates 8 (Figure 5) were prepared but they did not exhibit mesomorphism [53]. The Suzuki coupling reaction between tetrabromopyrene and appropriate boronic acid ester yields liquid crystalline tetraaryl-pyrenes 9 (Figure 5) [54, 55]. Similarly, Sonogashira-Hagihara reaction in between tetrabromopyrene 2 and 3,4,5-tridodecyloxyphenyl-acetylene yields 1,3,6,8-tetrakis(3,4,5-trisdodecyloxyphenylethynyl) pyrene

10. All these compounds display columnar mesophases [56]. Thus, the space filling around the core in inducing the mesomorphism is very important. An ambipolar charge carrier mobility of the order of 10^{-3} cm^2 V^{-1}s^{-1} was observed in discotic pyrene derivatives [57]. Compound 10 with ethynyl linkage is a highly fluorescent material with high quantum yield [58].

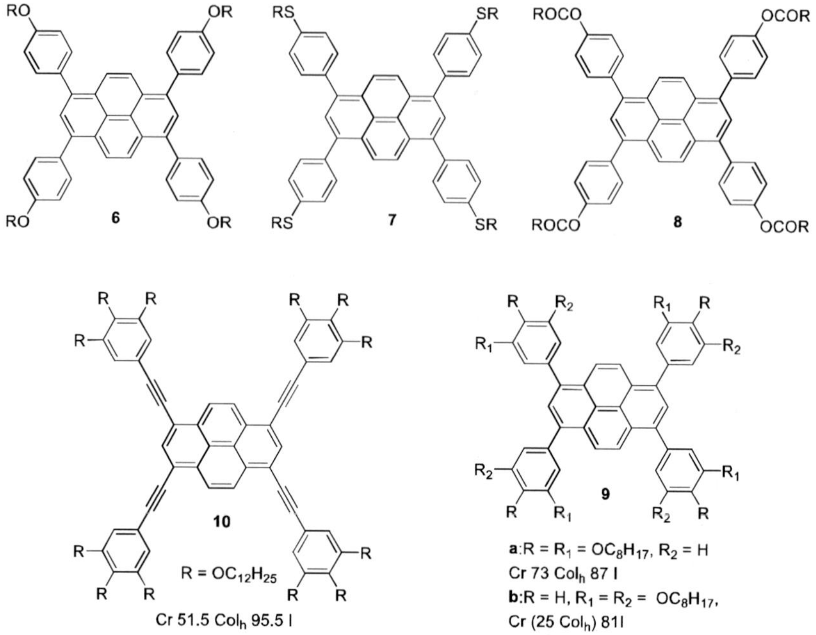

Figure 5. Molecular structures of tetraaryl and alkynylpyrene derivatives.

Pyrene hydrocarbon upon oxidation with sodium dichromate in sulphuric acid results a mixture of anti- and syn-type pyrene derivatives; pyrene-1,6-dione 11 and pyrene-1,10-dione 12 in equal amount (Figure 6) [59]. This mixture on treatment with an alcohol in presence of iron chloride yields corresponding alkoxypyrene derivatives 13 and 14. Compounds with longer alkoxy chain derivatives 13c (I 208 D$_L$ 184 Cr), 13d (I 184 D$_L$ 74 Cr), 14b (I 170 D$_L$ 134 Cr), 14c (I 148 D$_L$ 106 Cr) and 14d (I 139 D$_L$ 105 Cr) display discotic lamellar phases. Short chain derivatives, 13a, 13b and 14a were nonmesomorphic.

The monosubstituted pyrene dendrimers 15, 16, 17, 18 (Figure 7) were reported to exhibit hexagonal columnar phases [60-63]. The dendritic H-bonded molecules with bulky groups were previously reported to display columnar cubic mesophases [64, 65].

The liquid crystalline nature of 1,6-disubstituted pyrene 19 (Figure 8) was reported by Sagara and Kato [66]. It was prepared from 1,6-diethynylpyrene and the fan-shaped dendron via Sonogashira coupling. This compound exhibits an optically isotropic cubic phase from -35ºC to 175ºC.

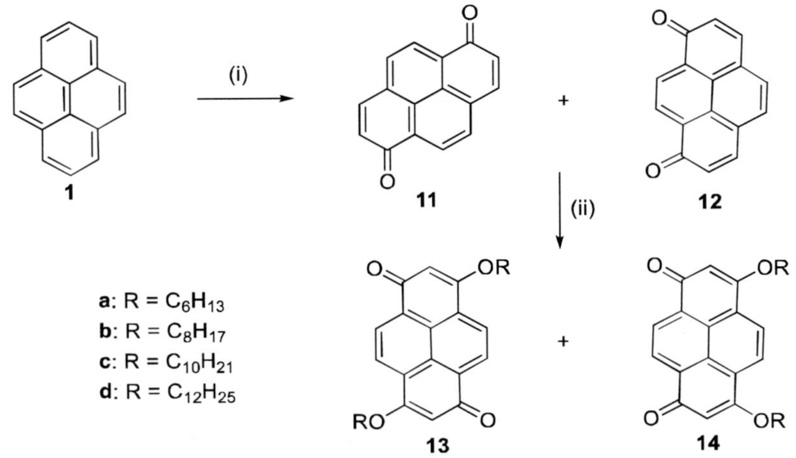
(i)
1
11
12
+
(ii)
a: R = C6H13
b: R = C8H17
c: R = C10H21
d: R = C12H25
13
14
OR
RO

Figure 6. Synthesis of dialkoxy syn- and anti-pyrenedione. (i) $Na_2Cr_2O_7$, H_2SO_4, (ii) ROH, $FeCl_3$.

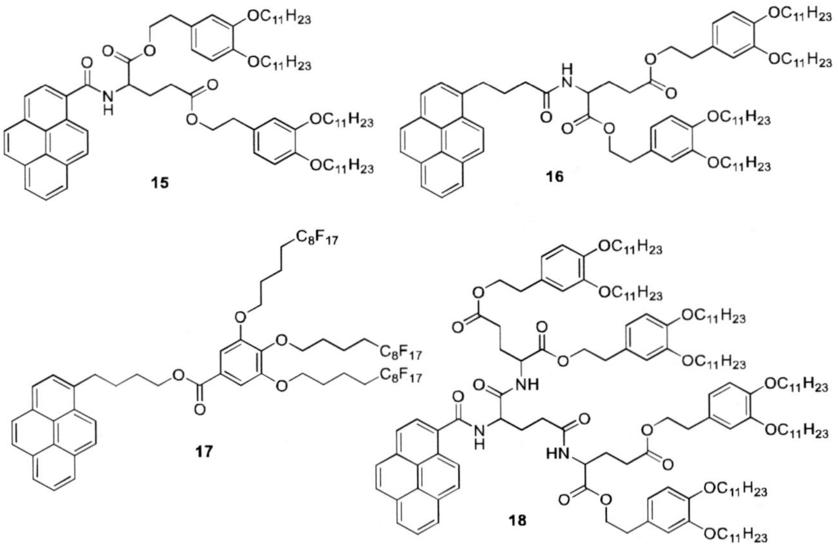
OC11H23
15
16
17
18
C8F17
NH
HN

Figure 7. Monosubstituted pyrene functionalized dendrimer liquid crystals.

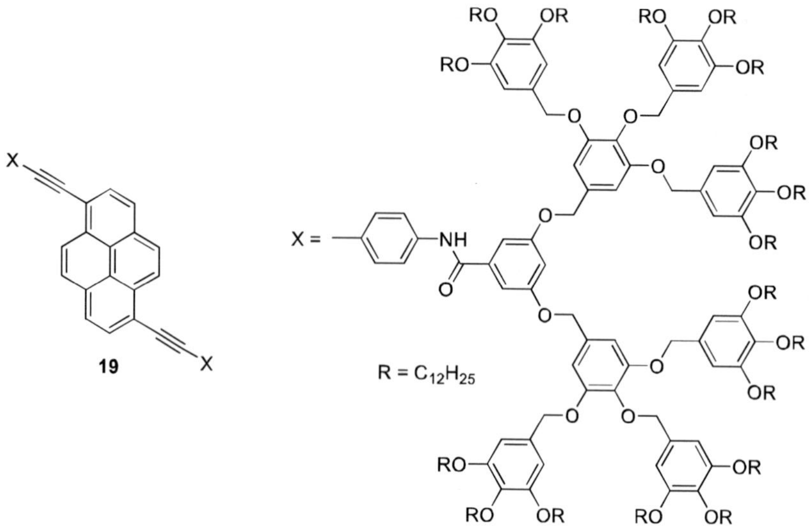

Figure 8. Chemical structure of disubstituted pyrene dendrimer.

Tchebotareva et al. prepared a hexabenzocoronene (HBC) derivative having a pyrene moiety attached covalently to it (Figure 9). The HBC-pyrene dyad 20 exhibits an ordered columnar phase with dramatically lowered isotropization temperature compared to parent HBC discotic [67]. At solid-liquid interface this dyad displays two-dimensional crystalline monolayer with uniform nanoscale segregation of the large and small aromatic systems as revealed by STM (Figure 10).

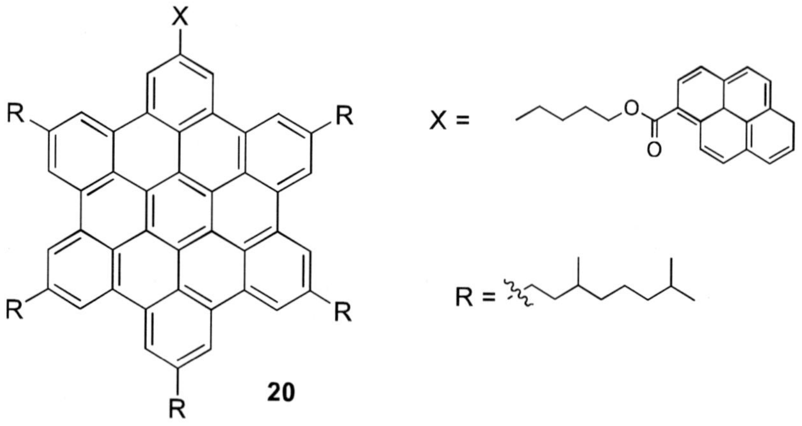

Figure 9. Hexabenzocoronene with pyrene unit.

Self-Assembling Supramolecular Structures of Pyrene 105

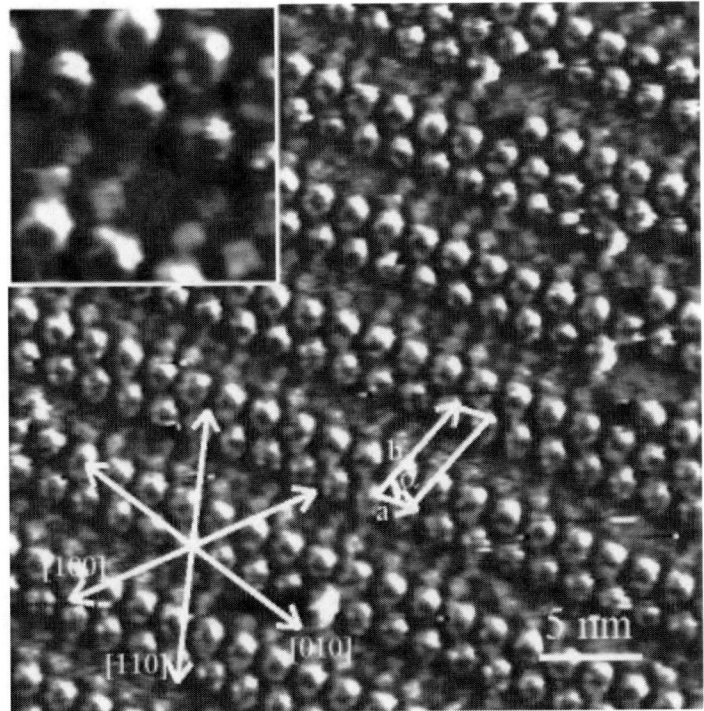

Figure 10. STM current image of 1 at the solid-liquid interface. Dimer row structures with smaller and darker bright spots between the dimer gaps. Inset: zoom-in image. Bias voltage = 1.2 V (tip positive); average tunnelling current = 100 pA; scan rate = 30 line/s (Reproduced with permission from Ref. 67; *J. Am. Chem. Soc.* 2003, 125, 9734-9739, Copyright {2003} American Chemical Society).

Four pyrene-based discotic mesogens with trialkylsilylethynyl units in the periphery 21 (Figure 11) were prepared by Hirose et al. [68]. Shorter chain derivatives 21a and 21b exhibit columnar mesophases. The charge carrier transport mobility (hole mobility), determined by the time-of-flight (TOF) method, in the columnar phases of these compounds was of the order of 10^{-2} cm^{2}V^{-1} s^{-1}. The high solubility of these discotics in various solvents and high charge mobility make these materials interesting candidates for optoelectronic devices.

On the other hand, 1,3,6,8-tetrakis((trimethysilyl)ethynyl)pyrene **22** (Figure 12) was reported to be non-liquid crystalline but exhibits columnar packing in the solid state. The powder XRD displays typical peaks associated with an ordered columnar phase but no fluidity was observed even at higher temperature [69].

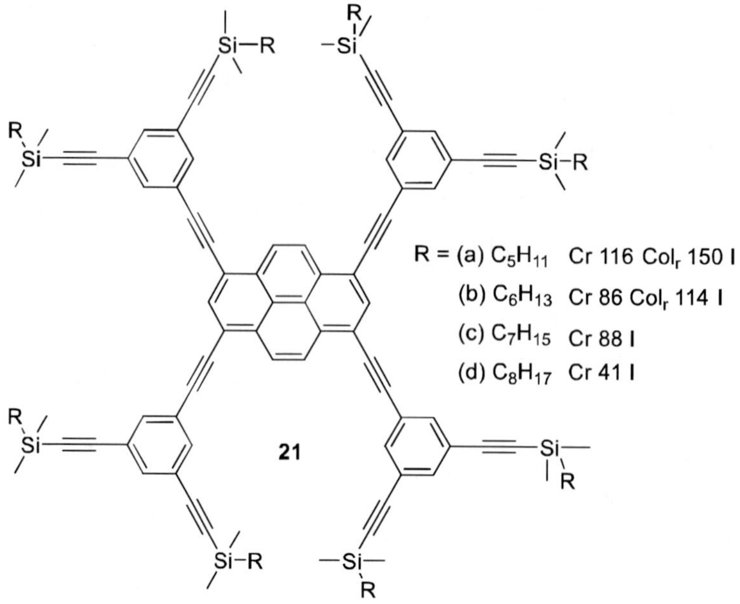

Figure 11. Tetrasubstituted pyrene derivatives with trialkylsilylethynyl units.

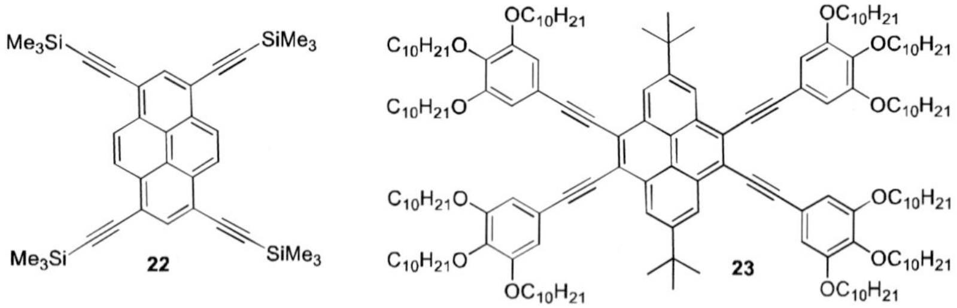

R = (a) C₅H₁₁ Cr 116 Col_r 150 I
(b) C₆H₁₃ Cr 86 Col_r 114 I
(c) C₇H₁₅ Cr 88 I
(d) C₈H₁₇ Cr 41 I

Figure 12. Tetrasubstituted pyrene derivatives with trimethysilylethynyl and arylethynyl units.

Kapf et al. prepared eight alkyloxy modified 2,7-di-tert-butyl-4,5,9,10-tetra(arylethynyl)pyrenes [70]. The typical Sonogashira coupling between 2,7-di-tert-butyl-4,5,9,10-tetrabromopyrene and trialkoxy-phenylacetylenes in presence of catalytic amount of Pd(II), CuI and triphenylphosphine afforded the desired materials in good yield. One of these compounds 23 (Figure 12) display a monotropic discotic nematic phase in a narrow temperature range between 44–58°C as characterized from polarizing optical microscopy.

Metal-free and metallized pyrenocyanines (pyrene-fused tetra-azaporphyrins 24 (Figure 13) were synthesized from pyrene-4,5-dicarbonitrile [71]. Peripheral substitution with linear and branched alkoxy chains not only make these materials soluble in organic solvents but also induces columnar mesophases. These materials were reported to display columnar mesophases at higher temperature but detailed mesophase transitions are not revealed.

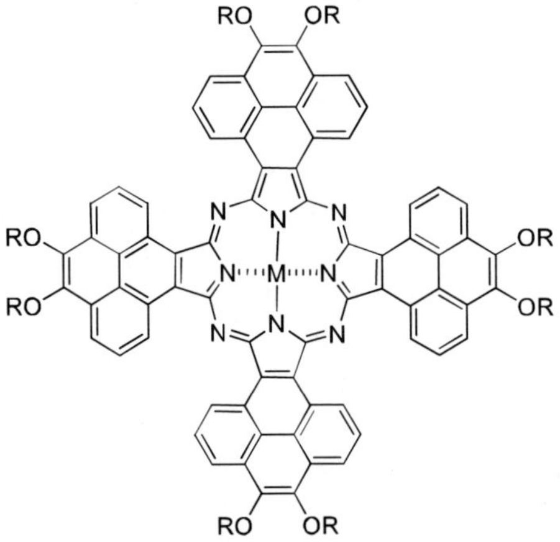

24a: M = Zn, R = C_8H_{17}
24b: M = H2, R = $C_{12}H_{25}$
24c: M = Zn, R = $C_{12}H_{25}$
24d: M = Pb, R = $C_{12}H_{25}$
24e: M = Zn, R = $C_{6,10}H_{33}$

Figure 13. Liquid crystalline pyrenocyanines.

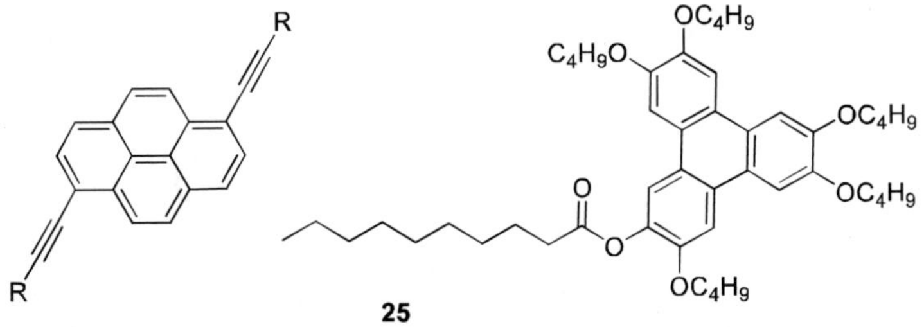

Figure 14. Structure of triphenylene-coupled pyrene derivative.

Triphenylene core is the most widely explored core in the field of DLCs. More than 1000 different DLCs have been realized from this core [72-74]. Zhang et al. prepared a discotic trimer in which two triphenylene mesogens were connected to a pyrene core 25 via alkyne linkages using Sonogashira coupling (Figure 14). Scanning tunnelling microscopy results show that the trimer assembled into a stable long-ranged zigzag structure on highly oriented pyrolytic graphite (HOPG) surface. However, surprisingly, no thermal behaviour of this interesting compound was disclosed [75].

Anetai et al. prepared fluorescent hydrogen-bonded pyrene derivative 26 (Figure 15) [76]. The alkylamide-substituted pyrene, N,N′,N″,N‴-tetra-(tetradecyl)-1,3,6,8-pyrenetetracarboxamide prepared from 1,3,6,8-tetrabromopyrene, exhibits hexagonal columnar mesophase above 295 K due to hydrogen bonding and π-stacking interactions. However, the transition temperature from columnar mesophase to the isotropic phase could not be observed due to thermal decomposition of the material. The compound was found to form fluorescent organogels (Figure 16) in various solvents. The yellow organogel was observed to emit green/blue fluorescence under UV light. The formation of three-dimensional entangled fibrous molecular assemblies was observed by (SEM) images of the xerogels. The cyclohexane gel exhibits lyotropic liquid crystalline phase. The intermolecular N−H···O=C hydrogen-bonded chains were reported to have two types of orientation along the π-stacking direction; one with dipole down and the other with dipole up (Figure 16). The pyrene derivative is reported to exhibit ferroelectric switching in the Col$_h$ phase due to the collective inversion of the intermolecular hydrogen-bonded chain. Further, it was observed that the electron transport properties in the Col$_h$ phase are effected by the macrodipole moment in the ferroelectric state.

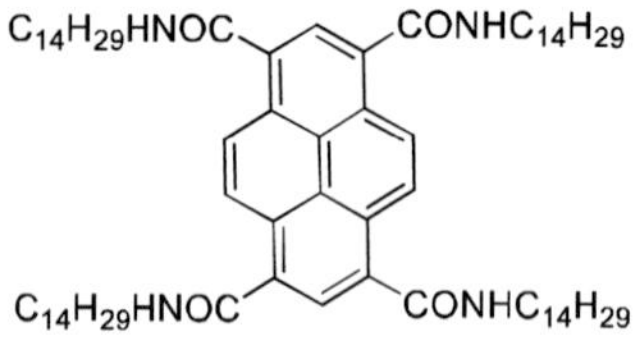

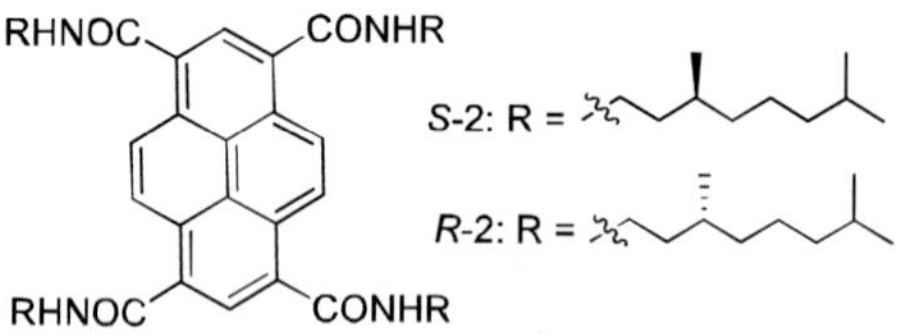

Figure 15. Hydrogen-bonded pyrene derivatives.

On the other hand, similar pyrene amides bearing chiral 3,7-dimethyl-1-octhylamide chains (S-2 and R-2) (Figure 15) did not show mesomorphism and ferroelectricity [77]. Nanoscale effects in one-dimensional columnar supramolecular ferroelectrics were observed by mixing S-2 with tetra-(tetradecyl)-1,3,6,8-pyrenetetracarboxamide.

A block copolymer 28 having a terminal pyrene unit (Figure 17) was synthesized using Grignard metathesis polymerization [78]. The discotic liquid crystalline mesogen, 6-(pyren-1-yloxy)hexyl methacrylate (PyMA), comprises a block that is attached to regioregular poly(3- hexylthiophene) (rr-

P3HT). The polymer exhibits melting transition at 198°C and crystallization at 166°C in the heating and cooling cycle. The glass transition of the methacrylate block was observed at 77°C. A weak transition at about 125°C in the cooling cycle was assigned to the liquid crystal mesophase transition of the pyrene. An organic field-effect transistor (OFET) in bottom-gate, top-contact geometry was constructed and its field-effect mobility in the order of 10^{-2} cm^2 V^{-1} s^{-1} were observed.

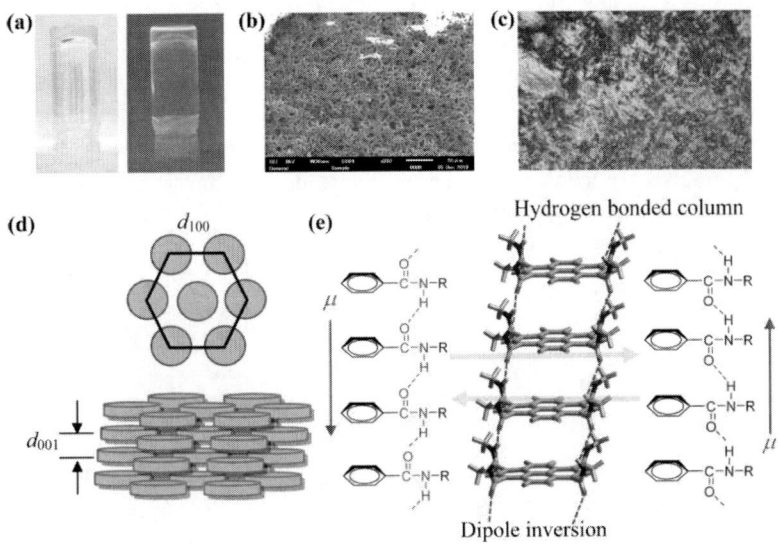

Figure 16. Diverse molecular assemblies of 1. (a) Organogels in hexane under visible light (left), and UV light (right), at 365 nm. (b) SEM image of the xerogel on HOPG (scale =50 μm). (c) POM image of the Colh mesophase at 380 K. (d) Schematic illustration of Colh mesophase with d100 and d001 spacing. (e) Hydrogen-bonded columnar structure and dipole inversion through the collective conformational change of the hydrogen-bonded chain along the π-stacking direction (Reproduced with permission from Ref. 76; *J. Phys. Chem. Lett.* 2015, 6, 1813).

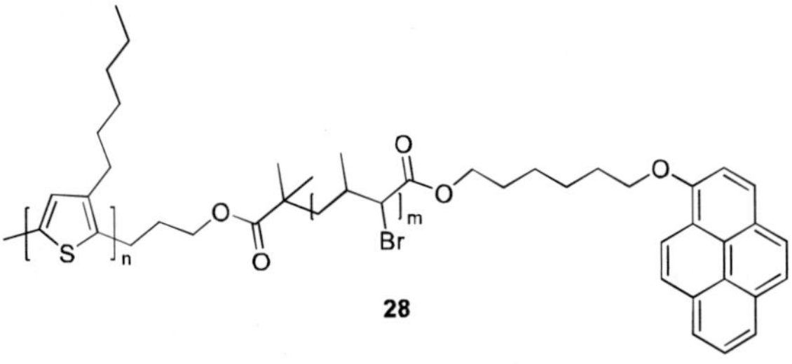

Figure 17. Block copolymer with a terminal pyrene unit.

The confinement of DLCs in nanoporous materials is an interesting strategy to prepare nanofibers. Such nanofibers are useful in organic electronics and, therefore, the confinement of DLCs in nanoporous materials has been extensively studied [79-83]. The DLC used in all these studies was pyrene-1,3,6,8-tetracarboxylic rac-2-ethylhexyl ester (Py4CEH) 5c (Figure 4). Anodized aluminum oxide (AAO) nanoporous membranes and nanoporous silica membranes were used to incorporate DLC. The DLC was confined into nanoporous samples by spontaneous imbibition in the liquid phase. Small-angle neutron scattering, X-ray diffraction studies, dielectric spectroscopy, differential scanning calorimetry, and theoretical studies were used to see the effects of confinement. "A homeotropic anchoring (face-on orientation of the disk-shape molecules at the interface) is favoured in all smooth cylindrical nanochannels of porous alumina while surface roughness of porous silicon promotes more disordered structures. Confinement in AAO membranes induces a large melting depression of the Col_h–Iso transition and the appearance of a hysteretic character" [79].

Conclusion

In this chapter, pyrene based self-assembling supramolecular compounds are presented. A number of pyrene derivatives form discotic liquid crystals upon appropriate substitution. First, a brief description of discotic liquid crystals is presented followed by a brief description of pyrene in the introduction part of the article. A large number of 1,3,6,8-substituted pyrene derivatives display mesomorphism. The synthesis, thermal behaviour and physical properties of these compounds are presented. Mesomorphic properties of some di-substituted pyrene are also described. Because of their interesting opto-electronic properties, these materials have huge potential in various device applications.

References

[1] Kumar S. Self-organization of disc-like molecules: Chemical aspects. *Chem Soc Rev*, 2006; 35:83-109.

[2] Balagurusamy VSK, Prasad SK, Chandrasekhar S, Kumar S. Quasi-one dimensional electrical conductivity and thermoelectric power studies on a discotic liquid crystal. *Pramana*, 1999; 53:3-11.

[3] Kumar S. *Chemistry of discotic liquid crystals: From monomers to polymers*. Boca Raton, (FL): CRC press; 2011.
[4] Bushby RJ, Lozman OR. Photoconducting liquid crystals. *Curr Opi Sol State Mat Sci*, 2002; 6:569-578.
[5] Sergeyev S, Pisula W, Geerts YH. Discotic liquid crystals: A new generation of organic semiconductors. *Chem Soc Rev*, 2007; 36:1902-1929.
[6] Pisula W, Feng X, Mullen K. Charge-carrier transporting graphene-type molecules. *Chem Mater*, 2011; 23: 554-567.
[7] Ozaki M, Yoneya M, Shimizu Y, Fujii, A. Carrier transport and device applications of the organic semiconductor based on liquid crystalline non-peripheral octaalkyl phthalocyanine. *Liq Cryst*, 2018; 45:2376-2389.
[8] Sahamir SR, Said SM, Sabri MFM, Mahmood MS, Kamarudin MAB, Moutanabbir O. Studies on relation between columnar order and electrical conductivity in HAT6 discotic liquid crystals using temperature-dependent Raman spectroscopy and DFT calculations. *Liq Cryst*, 2018; 45:522-535.
[9] Iino H, Hanna J-I, Haarer D, Bushby RJ. Fast electron transport in discotic columnar phases of triphenylene derivatives. *Jpn J Appl Phys*, 2006; 45:430.
[10] van de Craats AM, Warman JM, de Haas MP, Adam D, Simmerer J, Haarer D, Schuhmacher P. The mobility of charge carriers in all four phases of the columnar discotic material hexakis (hexylthio) triphenylene: combined TOF and PR-TRMC results. *Adv Mater*, 1996; 8:823-6.
[11] Kaafarani BR. Discotic liquid crystals for opto-electronic applications. *Chem Mater*, 2011; 23:378-96.
[12] Kumar M, Kumar S. Liquid crystals in photovoltaics: a new generation of organic photovoltaics. *Polym J*, 2017; 49:85-111.
[13] Pisula W, Feng X, Müllen K. Charge-carrier transporting graphene-type molecules. *Chem Mater*, 2011; 23:554-67.
[14] Wöhrle T, Wurzbach I, Kirres J, Kostidou A, Kapernaum N, Litterscheidt J, Haenle JC, Staffeld P, Baro A, Giesselmann F, Laschat S. Discotic liquid crystals. *Chem Rev*, 2016; 116:1139-241.
[15] Bisoyi HK, Kumar S. Liquid-crystal nanoscience: an emerging avenue of soft self-assembly. *Chem Soc Rev*, 2011; 40:306-19.
[16] Boden N, Bushby R, Clements J. Electron transport along molecular stacks in discotic liquid crystals. *J Mater Sci Mater Electron*, 1994; 5:83-8.
[17] Wu J, Pisula W, Müllen K. Graphenes as potential material for electronics. *Chem Rev*, 2007; 107:718-47.
[18] Hanna J-I. Charge carrier transport in liquid crystalline semiconductors. In: *Liquid crystalline semiconductors*. Springer; Dordrecht; 2013, pp. 39-64.
[19] Ohta K, Hatsusaka K, Sugibayashi M, Ariyoshi M, Ban K, Maeda F, Naito R, Nishizawa K, Van de Craats AM and Warman JM. Discotic liquid crystalline semiconductors. *Mol Cryst Liq Crys*, 2003; 397:25-45.
[20] Kumar M., Varshney S, Kumar S. Emerging nanoscience with discotic liquid crystals, *Polymer Journal*, (2021) 53:283-297.

[21] Kumar S. Investigations on discotic liquid crystals, *Liquid Crystals*, 2020, 47:8, 1195-1203.
[22] Bisoyi HK, Kumar S. Carbon-based liquid crystals: Art and science, *Liq Cryst*, 2011; 38:1427-49.
[23] Kumar S. Discotic liquid crystal-nanoparticle hybrid systems. *NPG Asia Mater*, 2014; 6:e82.
[24] Kumar S. Nanoparticles in the supramolecular order of discotic liquid crystals. *Liq Cryst*, 2014; 41:353-67.
[25] Bisoyi HK, Kumar S. Discotic nematic liquid crystals: science and technology. *Chem Soc Rev*, 2010; 39:264-85.
[26] Gowda, A., Kumar, S. Recent advances in discotic liquid crystal-assisted nanoparticles, *Materials*, 11, 382, 2018.
[27] Scott LT. *Chem Soc Rev*, 2015, 44, 6464-6471.
[28] Harvey R. *Curr Org Chem*, 2004, 8, 303-323.
[29] Cromwell A, Hager JJ. *J Am Chem Soc*, 1936, 58, 1051-1052.
[30] Weitzenbock R. *Monatshefte for Chemie*, 1913, 34, 193-223.
[31] Figueira-DuarteTM, Müllen K. *Chem Rev*, 2011, 111, 7260-7314.
[32] Yamana K, Iwai T. Ohtani Y, Sato S, Nakamura M, Nakano H. *Bioconjug Chem*, 2002, 13, 1266-1273.
[33] Okamoto A, Kanatani K, Saito I. *J Am Chem Soc*, 2004, 126, 4820-4827.
[34] Langenegger SM, Häner R. *Chem Commun*, 2004, 2792-3.
[35] Pu L. *Chem Rev*, 2004, 104, 1687-1716.
[36] Martinez-Manez R, Sancenon F. *Chem Rev*, 2003, 103, 4419-4476.
[37] Daub J, Engl R, Kurzawa J, Miller SE, Schneider S, Stockmann A, Wasielewski MR. *J Phys Chem A*, 2001, 105, 5655-5665.
[38] Baker LA, Crooks RM. *Macromolecules*, 2000, 33, 9034-9039.
[39] Modrakowski C, Flores SC, Beinhoff M, Schluter AD. *Synthesis*, 2001, 2143-2155.
[40] Chaiken RF, Kearns DR. *J Chem Phys*, 1968, 49, 2846-2850.
[41] Holroyd RA, Preses JM, Boettcher EH, Schmidt WF. *J Phys Chem*, 1984, 88, 744-749.
[42] Jones IIG, Vullev VI. *Org Lett*, 2002, 4, 4001-4004.
[43] Yamana K, Fukunaga Y, Ohtani Y, Sato S, Nakamura M, Kim WJ, Akaike T, Maruyama A. *Chem Commun*, 2005, 2509-2511.
[44] Hwang GT, Seo YJ, Kim BH. *J Am Chem Soc*, 2004, 126, 6528-6529.
[45] Fujimoto K, Shimizu H, Inouye M. *J Org Chem*, 2004, 69, 3271-3275.
[46] Tomkiewicz Y, Weinreb A. *Chem Phys Lett*, 1969, 3, 229-230.
[47] Bock H, Helfrich W. *Liq Cryst*, 1995, 18, 707-713.
[48] Hirose T, Kawakami O, Yasutake M. *Mol Cryst Liq Cryst*, 2006, 451, 65-74.
[49] Hassheider T, Benning SA, Kitzerow H-S, Achard M-F, Bock H. *Angew Chemie Int Ed*, 2001, 40, 2060-2063.
[50] Keuker-Baumann S, Bock H, Della Sala F, Benning SA, Haßheider T, Frauenheim T, Kitzerow H-S. *Liq Cryst*, 2001, 28, 1105-1113.

[51] Dantras E, Dandurand J, Lacabanne C, Laffont L, Tarascon JM, Archambeau S, Seguy I, Destruel P, Bock H, Fouet S. *Phys Chem Chem Phys,* 2004, 6, 4167.
[52] Krause C, Yin H, Cerclier C, Morineau D, Wurm A, Schick C. Molecular dynamics of a discotic liquid crystal investigated by a combination of dielectric relaxation and specific heat spectroscopy, *Soft Matter,* 2012, 8, 11115.
[53] de Halleux V, Calbert JP, Brocorens P, Cornil J, Declercq JP, Bredas JL, Geerts Y *Adv Fun Mater,* 2004, 14, 649-659.
[54] Sienkowska MJ, Monobe H, Kaszynski P, Shimizu Y. *J Mater Chem,* 2007, 17, 1392-1398.
[55] Sienkowska MJ, Farrar JM, Zhang F, Kusuma S, Heiney PA, Kaszynski P. *J Mater Chem,* 2007, 17, 1399-1411.
[56] Hayer A, de Halleux V, Kohler A, El-Garoughy A, Meijer EW, Barbera J, Tant J, Levin J, Lehmann M, Gierschner J, Cornil J, Geerts YH. *J Phys Chem B,* 2006, 110, 7653-7659.
[57] Sienkowska MJ, Monobe H, Kaszynski P, Shimizu Y. *J Mater Chem,* 2007, 17, 1392.
[58] Hayer A, de Halleux V, Köhler A, El-Garoughy A, Meijer EW, Barberá J, Tant J, Levin J, Lehmann M, Gierschner J. *J Phys Chem B,* 2006, 110, 7653-7659.
[59] Yasutake M, Fujihara T, Nagasawa A, Moriya K, Hirose T. *European J Org Chem,* 2008, 2008, 4120-4125.
[60] Kim YH, Yoon DK, Lee EH, Ko YK, Jung H-T. *J Phys Chem B,* 2006, 110, 20836-20842.
[61] Percec V, Glodde M, Bera TK, Miura Y, Shiyanovskaya I, Singer KD, Balagurusamy VSK, Heiney PA, Schnell I, Rapp A. *Nature,* 2002, 419, 384-387.
[62] Shiyanovskaya I, Singer KD, Percec V, Bera TK, Miura Y, Glodde M. *Phys Rev B,* 2003, 67, 035204.
[63] Kamikawa Y, Kato T. *Langmuir,* 2007, 23, 274-278.
[64] Kato T, Matsuoka T, Nishii M, Kamikawa Y, Kanie K, Nishimura T, Yashima E, Ujiie S. *Angew Chemie Int Ed,* 2004, 43, 1969-1972.
[65] Kamikawa Y, Nishii M, Kato T. *Chem - A Eur J,* 2004, 10, 5942-5951.
[66] Sagara Y, Kato T. *Angew Chemie Int Ed,* 2008, 47, 5175-5178.
[67] Tchebotareva N, Yin X, Watson MD, Samori P, Rabe JP, Mullen K. *J Am Chem Soc,* 2003, 125, 9734-9739.
[68] Takuji Hirose, Yuki Shibano, Yutaro Miyazaki, Norihito Sogoshi, Seiichiro Nakabayashi and Mikio Yasutake. Synthesis and Hole Transport Properties of Highly Soluble Pyrene-Based Discotic Liquid Crystals with Trialkylsilylethynyl Groups, *Molecular Crystals and Liquid Crystals,* 2011, 534:1, 81-92.
[69] Feng Xu, Takanori Nishida, Kenta Shinohara, Lifen Peng, Makoto Takezaki, Takahiro Kamada, Haruo Akashi, Hiromu Nakamura, Kouki Sugiyama, Kazuchika Ohta, Akihiro Orita and Junzo Otera. Trimethylsilyl Group Assisted Stimuli Response: Self-Assembly of 1,3,6,8-Tetrakis((trimethysilyl)ethynyl)pyrene, *Organometallics,* 2017, 36, 556-563.

[70] Kapf Andreas, Eslahi Hassa, Blanke Meik, Saccone Marco, Giese Michael, Albrecht Marcel. Alkyloxy modified pyrene fluorophores with tunable photophysical and crystalline properties, *New J Chem,* 2019, 43, 6361.

[71] Hayato Anetai, Kohei Sambe, Takashi Takeda, Norihisa Hoshino and Akutagawa T. Nanoscale Effects in One-Dimensional Columnar Supramolecular Ferroelectrics, *Chem Eur J,* 2019, 25, 11233-11239.

[72] Kumar S. Recent developments in the chemistry of triphenylenebased discotic liquid crystals. *Liq Cryst,* 2004; 31:1037-59.

[73] Kumar S. Triphenylene-based discotic liquid crystal dimers, oligomers and polymers. *Liq Cryst,* 2005; 32:1089-1113.

[74] Pal SK, Setia S, Avinash BS, Kumar S. Triphenylene-based discotic liquid crystals: recent advances. *Liq Cryst,* 2013; 40:1769-1816.

[75] Zhang Xue-mei, Wang Hai-feng, Wang Shuai, Shen Yong-tao, Yang Yan-lian, Deng Ke, Zhao Ke-qing, Zeng Qing-dao, Wang Chen. Triphenylene Substituted Pyrene Derivative: Synthesis and Single Molecule Investigation, *J Phys Chem C,* 2013, 117, 307-312.

[76] Hayato Anetai, Yoshifumi Wada, Takashi Takeda, Norihisa Hoshino, Shunsuke Yamamoto, Masaya Mitsuishi, Taishi Takenobu and Tomoyuki Akutagawa. Fluorescent Ferroelectrics of Hydrogen-Bonded Pyrene Derivatives, *J Phys Chem Lett,* 2015, 6, 1813-1818.

[77] Hayato Anetai, Kohei Sambe, Takashi Takeda, Norihisa Hoshino and Tomoyuki Akutagawa. Nanoscale Effects in One-Dimensional Columnar Supramolecular Ferroelectrics, *Chem Eur J,* 2019, 25, 11233–11239.

[78] Pathiranage TMSK, Ma Z, Udamulle Gedara CM, Pan X, Lee Y, Gomez ED, Biewer MC, Matyjaszewski K, Stefan MC. Improved Self-Assembly of P3HT with Pyrene-Functionalized Methacrylates, *ACS Omega,* 2021, 6, 27325-27334.

[79] Cerclier CV, Ndao M, Busselez R, Lefort R, Grelet E, Huber P, Kityk AV, Noirez L, Schönhals A, Morineau D. Structure and Phase Behavior of a Discotic Columnar Liquid Crystal Confined in Nanochannels, *J Phys Chem C,* 2012, 116, 18990-18998.

[80] Krause C, Schönhals A. Phase Transitions and Molecular Mobility of a Discotic Liquid Crystal under Nanoscale Confinement, *J Phys Chem C,* 2013, 117, 19712-19720.

[81] Kityk AV, Busch M, Rau D, Calus S, Cerclier CV, Lefort R, Morineau D, Grelet E, Krause C, Schonhals A, Frick B, Huber P. Thermotropic orientational order of discotic liquid crystals in nanochannels: An optical polarimetry study and a Landau–de Gennes analysis, *Soft Matter,* 2014, 10, 4522.

[82] Ndao M, Lefort R, Cerclier CV, Busselez R, Morineau D, Frick B, Ollivier J, Kityk AV, Huber P. Molecular dynamics of pyrene based discotic liquid crystals confined in nanopores probed by incoherent quasielastic neutron scattering, *RSC Adv,* 2014, 4, 59358.

[83] Calus S, Kityk AV, Borowik L, Lefort R, Morineau D, Krause C, Schonhals A, Busch M, Huber P. High-resolution dielectric study reveals pore-size-dependent

orientational order of a discotic liquid crystal confined in tubular nanopores, *Phy Rev E,* 2015. 92, 012503.

Biographical Sketch

Sandeep Kumar

Affiliation: Nitte Meenakshi Institute of Technology, Yelahanka, Bangalore - 560064, India

Education: PhD

Business Address: Nitte Meenakshi Institute of Technology, Yelahanka, Bangalore - 560064, India

Research and Professional Experience: 42 years

Professional Appointments: Professor

Honors: E.T.S. Walton Visiting Professor Award, LG Philips Display Mid-Career Award

Publications from the Last 3 Years: 50

Bibliography

Ambient monitoring of benzo-a-pyrene (B[a]P) emissions at two oil refineries: energy institute research report
LCCN 2009396325
Type of material Book
Main title Ambient monitoring of benzo-a-pyrene (B[a]P) emissions at two oil refineries: Energy Institute research report.
Published/Created London: Energy Institute Press, [2008]
Description vi, 9 p.: ill.; 30 cm.
ISBN 9780852935064 (pbk.)
0852935064 (pbk.)
LC classification TD195.P4 A47 2008
Related names Energy Institute (Great Britain)
Great Britain. Environment Agency.
LC Subjects Petroleum refineries--Environmental aspects--Great Britain.
Benzopyrene.
Air--Pollution--Great Britain--Measurement.
Notes "Environment Agency" - Cover.
"July 2008".
Includes bibliographical references (p. 9).

Biomaterial applications: macro to nanoscales
LCCN 2014953120
Type of material Book
Main title Biomaterial applications: macro to nanoscales / edited by Sabu Thomas, PhD, Nandakumar Kalarikkal, PhD, Weimin Yang, MD, and Snigdha S. Babu.

Published/Produced	Toronto: Apple Academic Press, [2015]
Description	xviii, 213 pages: illustrations; 23 cm
ISBN	9781771880275 (bound)
	1771880279 (bound)
LC classification	TP248.65.P62 I57 2012
Related names	Thomas, Sabu, editor.
	Kalarikkal, Nandakumar, editor.
	Yang, Weimin, MD, editor.
	Babu, Snigdha S., editor.
	ICNP (Conference) (3rd: 2012: Kottayam, India)
Contents	1. Green organic-inorganic hybrid material from plant oil polyol / Eram Sharmin, Mudsser Azam, Fahmina Zafar, Deewan Akram, Qazi Mohd. Rizwanul Haq, and Sharif Ahmad - 2. Bio-hybrid 3D tubular scaffolds for vascular tissue engineering: a materials perspective / Harsh Patel, Roman Garcia, and Vinoy Thomas3. Polymers for use in the monitoring and treatment of waterborne protozoa / Helen Bridle and Moushumi Ghosh - 4. Synthesis of Polypyrrole/TiO2 nanoparticles in water by chemical oxidative polymerization / Yang Tan, Michel B. Johnson, and Khashayar Ghandi - 5. Poly (lactic acid) based hybrid composite films containing ultrasound treated cellulose and poly (ethylene glycol) as plasticizer and reaction media / Katalin Halász, Mandar P. Badve, and Levente Csóka - 6. An experimental observation of disparity in mechanical properties of turmeric fiber reinforced polyester composites / Nadendla srinivasababu, J. Suresh Kumar and K. Vijaya Kumar Reddy - 7. Wavelength dependence of SERS spectra of pyrene / F. Hubenthal, D. Blázquez Sánchez, R. Ossig, H. Schmidt, and H.-D. Kronfeldt - 8. Emerging therapeutic applications of bacterial exopolysaccharides / P. Priyanka, A.B. Arun, and P.D. Rekha - 9. Preparation and properties of composite films from modified cellulose fiber-reinforced with different polymers / Sandeep S. Laxmishwar and G.K. Nagaraja - 10. Natural bio

Bibliography 119

	resources: the unending source of nanofactory / Balaprasad Ankamwar.
LC Subjects	Biomedical materials.
	Biopolymers.
Other Subjects	Biocompatible Materials.
	Biopolymers.
	Biomedical materials.
	Biopolymers.
Form/Genre	Congresses.
Notes	Based on papers presented at the ICNP 2012, Third International Conference on Natural Polymers, Bio-Polymers, Bio-Materials, their Composites, Blends, IPNs, Polyelectrolytes and Gels: Macro to Nano Scales took place at Mahatma Gandhi University, Kottayam, Kerala, India, on October 26, 27, and 28, 2012.
	Includes bibliographical references and index.
Additional formats	Electronic version: 9781482252767

Bioremediation: applications for environmental protection and management

LCCN	2019763956
Type of material	Book
Main title	Bioremediation: Applications for Environmental Protection and Management / edited by Sunita J. Varjani, Avinash Kumar Agarwal, Edgard Gnansounou, Baskar Gurunathan.
Edition	1st ed. 2018.
Published/Produced	Singapore: Springer Singapore: Imprint: Springer, 2018.
Description	1 online resource (XVI, 411 pages 52 illustrations, 34 illustrations in color.)
	PDF
ISBN	9789811074851
Related names	Agarwal, Avinash Kumar, editor.
	Gnansounou, Edgard, editor.
	Gurunathan, Baskar, editor.
	Varjani, Sunita J, editor.

Summary

This book examines bioremediation technologies as a tool for environmental protection and management. It provides global perspectives on recent advances in the bioremediation of various environmental pollutants. Topics covered include comparative analysis of bio-gas electrification from anaerobic digesters, mathematical modeling in bioremediation, the evaluation of next-generation sequencing technologies for environmental monitoring in wastewater abatement; and the impact of diverse wastewater remediation techniques such as the use of nanofibers, microbes and genetically modified organisms; bioelectrochemical treatment; phytoremediation; and biosorption strategies. The book is targeted at scientists and researchers working in the field of bioremediation.

Contents

Introduction to Environmental Protection and Management - Mathematical Modeling in bioremediation - Evaluation of Next-Generation sequencing technologies for environmental monitoring in waste water abatement - Genetically modified organisms and its impact on the enhancement bioremediation - Integration of lignin removal from black liquor and biotransformation process - Role of Nanofibers in bioremediation - Bioremediation of industrial wastewater using bioelectrochemical treatment - Biosorption strategies in the remediation of toxic pollutants from contaminated water bodies - Bioremediation of Heavy Metals - Pesticides Bioremediation - Application of microbes in remediation of hazardous wastes: A review - Phytoremediation techniques for the removal of dye in waste water - Phenol degradation from industrial wastewater by engineered microbes - Insect gut bacteria and their potential application in degradation of lignocellulosic biomass: a review - Bioremediation of volatile organic compounds in biofilters - Bioremediation of industrial and municipal waste water using

	microalgae - Phytoremediation of Textile Dye Effluents - Role of biosurfactants in enhancing the microbial degradation of pyrene - Bioremediation of nitrate contaminated wastewater and soil.
LC Subjects	Biochemical engineering. Biotechnology. Environmental engineering. Pollution. Water pollution. Water quality.
Other Subjects	Environmental Engineering/Biotechnology. Biochemical Engineering. Pollution, general. Water Quality/Water Pollution.
Additional formats	Print version: Bioremediation. 9789811074844 (DLC) 2017959893 Printed edition: 9789811074844 Printed edition: 9789811074868 Printed edition: 9789811356421
Series	Energy, Environment, and Sustainability, 2522-8366 Energy, Environment, and Sustainability, 2522-8366

DNA adducts: formation, detection and mutagenesis

LCCN	2009050733
Type of material	Book
Main title	DNA adducts: formation, detection and mutagenesis / Emerson Álvarez and Roberto Cunha, editors.
Published/Created	New York: Nova Biomedical Books, c2010.
Description	xiii, 232 p.: ill. (some col.); 27 cm.
ISBN	9781607414339 (hardcover) 1607414333 (hardcover)
LC classification	QP624.74 .D63 2010
Related names	Álvarez, Emerson. Cunha, Roberto.
Contents	Tidying up a molecular catastrophe: in vitro and in vivo repair of unique DNA-protein cross links / David J. Baker, Arthur D. Riggs, Timothy R. O'Connor - In vivo mutagenic effects of alkylating agents eliciting different DNA-adducts / Henriqueta

	Louro, Maria João Silva - DNA adduct formation and genotoxicity assessment in plants / Frédérique Weber-Lotfi ... [et al.] - Diet, life-style habits and malondialdehyde-deoxyguanosine adducts in a group of subjects resident in the Rayong Province, Thailand / M. Peluso ... [et al.] - Drug-DNA adducts formed by formaldehyde activation of anthracyclines and related anti-cancer agents / Don R. Phillips ... [et al.] - Differential modulation of benzo(a)pyrene-derived DNA adducts in MCF-7 cells by marine compounds / Jamal M. Arif ... [et al.] - Potency of air pollutants at DNA adduct formation and assessment by in vivo mutagenesis / Yasunobu Aoki ... [et al.] - GSTM1 and XRCC3 polymorphisms: effects on the levels of aflatoxin B1-DNA adducts / Xi-Dai Long ... [et al.] - Oxidative DNA damage and DNA adducts induced by polycyclic aromatic hydrocarbons and aromatic amines / Shosuke Kawanishi and Shiho Ohnishi - Usefulness of bulky DNA adduct formation as a biological marker of exposure to airborne particulate matter (PM2.5) in in vitro cell lung models: a comparative study / Sylvain Billet ... [et al.].
LC Subjects	DNA adducts.
	Genetic toxicology.
Other Subjects	DNA Adducts.
Notes	Includes bibliographical references and index.
Series	DNA: properties and modifications, functions and interactions, recombination and applications series
	DNA--properties and modifications, functions and interactions, recombination and applications series.

DNA damage detection in situ, ex vivo, and in vivo: methods and protocols

LCCN	2010938360
Type of material	Book
Main title	DNA damage detection in situ, ex vivo, and in vivo: methods and protocols / edited by Vladimir V. Didenko.
Published/Created	New York; London: Humana Press, c2011.

Description	xiii, 313 p.: ill. (some col.); 27 cm.
ISBN	9781603274081 (hbk.: alk. paper)
	1603274081 (hbk.: alk. paper)
LC classification	QH465.A1 D485 2011
Related names	Didenko, Vladimir V.
Contents	In situ detection of apoptosis by the TUNEL assay: an overview of techniques / Deryk T. Loo - Combination of TUNEL assay with immunohistochemistry for simultaneous detection of DNA fragmentation and oxidative cell damage / Alexander E. Kalyuzhny - EM-ISEL: a useful tool to visualize DNA damage at the ultrastructural level / Antonio Migheli - In situ labeling of DNA breaks and apoptosis by T7 DNA polymerase / Vladimir V. Didenko - In situ ligation: a decade and a half of experience / Peter J. Hornsby and Vladimir V. Didenko - In situ ligation simplified: using PCR fragments for detection of double-strand DNA breaks in tissue sections / Vladimir V. Didenko - 5'OH DNA breaks in apoptosis and their labeling by topoisomerase-based approach / Vladimir V. Didenko - Detection of DNA strand breaks in apoptotic cells by flow- and image-cytometry / Zbigniew Darzynkiewicz and Hong Zhao - Fluorochrome-labeled inhibitors of caspases: convenient in vitro and in vivo markers of apoptotic cells for cytometric analysis / Zbigniew Darzynkiewicz ... [et al.] - Combining fluorescent in situ hybridization with the comet assay for targeted examination of DNA damage and repair / Sergey Shaposhnikov, Preben D. Thomsen, and Andrew R. Collins - Simultaneous labeling of single- and double-strand DNA breaks by DNA breakage detection-FISH (DBD-FISH) / José Luis Fernández, Dioleyda Cajigal, and Jaime Gosálvez - Co-localization of DNA repair proteins with UV-induced DNA damage in locally irradiated cells / Jennifer Guerrero-Santoro, Arthur S. Levine, and Vesna Rapić-Otrin - Ultrasound imaging of apoptosis:

spectroscopic detection of DNA-damage effects at high and low frequencies / Roxana M. Vlad, Michael C. Kolios, and Gregory J. Czarnota - Quantifying etheno-DNA adducts in human tissues, white blood cells, and urine by ultrasensitive ^{32}P-postlabeling and immunohistochemistry / Jagadeesan Nair ... [et al.] - ELISpot assay as a tool to study oxidative stress in peripheral blood mononuclear cells / Jodi Hagen, Jeffrey P. Houchins, and Alexander E. Kalyuzhny - Cytokinesis-block micronucleus cytome assay in lymphocytes / Philip Thomas and Michael Fenech - Buccal micronucleus cytome assay / Philip Thomas and Michael Fenech - [Gamma]-H2AX detection in peripheral blood lymphocytes, splenocytes, bone marrow, xenografts, and skin / Christophe E. Redon ... [et al.] - Immunologic detection of benzo(a)pyrene-DNA adducts / Regina M. Santella and Yu-Jing Zhang - Non-invasive assessment of oxidatively damaged DNA: liquid chromatography-tandem mass spectrometry analysis of urinary 8-oxo-7,8-dihydro-2'-deoxyguanosine / Vilas Mistry ... [et al.] - Assessing sperm DNA fragmentation with the sperm chromatin dispersion test / José Luis Fernández ... [et al.]

LC Subjects	DNA damage.
	DNA damage--Research--Methodology.
	DNA--Analysis.
Other Subjects	DNA Damage--Laboratory Manuals.
	DNA--analysis--Laboratory Manuals.
	Microbiological Techniques--Laboratory Manuals.
Notes	Includes bibliographical references and index.
Additional formats	Also available online.
	Online version: DNA damage detection in situ, ex vivo, and in vivo. New York, N.Y.: Humana Press, c2011 9781603274098 (OCoLC)681767573
Series	Methods in molecular biology; v. 682
	Springer protocols
	Methods in molecular biology (Clifton, N.J.); v. 682.
	Springer protocols.

Evaluating evidence of mechanisms in medicine: principles and procedures

LCCN	2019744377
Type of material	Book
Personal name	Parkkinen, Veli-Pekka. author.
Main title	Evaluating Evidence of Mechanisms in Medicine: Principles and Procedures / by Veli-Pekka Parkkinen, Christian Wallmann, Michael Wilde, Brendan Clarke, Phyllis Illari, Michael P Kelly, Charles Norell, Federica Russo, Beth Shaw, Jon Williamson.
Edition	1st ed. 2018.
Published/Produced	Cham: Springer International Publishing: Imprint: Springer, 2018.
Description	1 online resource (XVIII, 125 pages 14 illustrations) PDF
Rights advisory	Creative Commons Attribution 4.0 International. CC BY 4.0 https://creativecommons.org/licenses/by/4.0/
Access advisory	Unrestricted online access
Links	https://hdl.loc.gov/loc.gdc/gdcebookspublic.2019744377
ISBN	9783319946108
Related names	Clarke, Brendan. author.
	Illari, Phyllis. author.
	Kelly, Michael P. author.
	Norell, Charles. author.
	Russo, Federica. author.
	Shaw, Beth. author.
	Wallmann, Christian. author.
	Wilde, Michael. author.
	Williamson, Jon. author.
Summary	This book is open access under a CC BY license. This book is the first to develop explicit methods for evaluating evidence of mechanisms in the field of medicine. It explains why it can be important to make this evidence explicit, and describes how to take such evidence into account in the evidence appraisal process. In addition, it develops procedures for seeking evidence of mechanisms, for evaluating evidence of mechanisms, and for combining this

evaluation with evidence of association in order to yield an overall assessment of effectiveness. Evidence-based medicine seeks to achieve improved health outcomes by making evidence explicit and by developing explicit methods for evaluating it. To date, evidence-based medicine has largely focused on evidence of association produced by clinical studies. As such, it has tended to overlook evidence of pathophysiological mechanisms and evidence of the mechanisms of action of interventions. The book offers a useful guide for all those whose work involves evaluating evidence in the health sciences, including those who need to determine the effectiveness of health interventions and those who need to ascertain the effects of environmental exposures.

Contents

1 Introduction - 1.1 What is a mechanism? - 1.2 Where does evidence of a mechanism come from? - 1.3 Why consider evidence of mechanisms? - 1.3.1 Evaluating efficacy - 1.3.2 Evaluating external validity - 1.3.3 Other questions - 1.3.4 Importance of considering evidence of mechanisms - 2 How to consider evidence of mechanisms: a summary - 2.1 Questions to address - 2.2 Quality level of evidence and status of claim - 2.3 Identifying evidence of mechanisms in the literature - 2.4 Evaluating evidence of mechanisms - 2.5 Using evidence of mechanisms to evaluate causal claims - 2.6 Overall approach - 3 Identifying evidence of mechanisms in the literature - 3.1 Hypothesize a mechanism - 3.2 Search the literature - 3.3 Identify the evidence most relevant to the mechanism hypothesis - 3.4 Presenting the evidence of mechanisms - 4 Evaluating evidence of mechanisms - 4.1 Considerations for evaluating evidence of mechanism - 4.2 Presenting quality of evidence of mechanisms - 5 Using evidence of mechanisms to evaluate efficacy and external validity - 5.1 Efficacy - 5.2 External validity - 6 Glossary - 7 References - 8 Acknowledgements - 9 Appendix A.

	A critical appraisal tool for evidence of mechanisms - 10 Appendix B. Grade tables with mechanism assessment - 11 Appendix C: Databases for evidence of mechanisms - 12 Appendix D: Assessing exposures - 12.1 Example: carcinogenicity of benzo[a]pyrene - 12.2 Comparison to IARC - 12.3 Molecular epidemiology - 12.4 Comparison to Syrina
LC Subjects	Epistemology.
	Medicine-Philosophy.
Other Subjects	Philosophy of Medicine.
	Epistemology.
Additional formats	Print version: Evaluating evidence of mechanisms in medicine. 9783319946092 (DLC) 2018945456
	Printed edition: 9783319946092
	Printed edition: 9783319946115
Series	SpringerBriefs in Philosophy, 2211-4548
	SpringerBriefs in Philosophy, 2211-4548

Flavian epic

LCCN	2015956478
Type of material	Book
Main title	Flavian epic / Edited by Antony Augoustakis.
Published/Produced	Oxford, United Kingdom: Oxford University Press, [2016]
Description	xii, 538 pages; 23 cm
ISBN	9780199650668
	0199650667
LC classification	PA6791.V4 F53 2016
Related names	Augoustakis, Antony, editor.
Summary	The epics of the three Flavian poets-Silius Italicus, Statius, and Valerius Flaccus-have, in recent times, attracted the attention of scholars, who have re-evaluated the particular merits of Flavian poetry as far more than imitation of the traditional norms and patterns. Drawn from sixty years of scholarship, this edited collection is the first volume to collate the most influential modern academic writings on Flavian epic poetry, revised and updated to provide both scholars and students alike with a broad yet comprehensive

Contents

overview of the field. A wide range of topics receive coverage, and analysis and interpretation of individual poems are integrated throughout. The plurality of the critical voices included in the volume presents a much-needed variety of approaches, which are used to tackle questions of intertextuality, gender, poetics, and the social and political context of the period. In doing so, the volume demonstrates that by engaging in a complex and challenging intertextual dialogue with their literary predecessors, the innovative epics of the Flavian poets respond to contemporary needs, expressing overt praise, or covert anxiety, towards imperial rule and the empire. Introduction: Flavian epic Renaissance (1990-2015) / Antony Augoustakis - Ratis audax: Valerius Flaccus' bold ship / Martha A. Davis - Virgilianisms in Valerius Flaccus: a contribution to the study of epic language in the imperial age / Roberta Nordera - The restoration of ancient models: epic tradition and mannerist technique in Valerius' Argonautica 6 / Marco Fucecchi - Allusion and narrative possibility in the Argonautica of Valerius Flaccus / Andrew Zissos - Parce metu, Cytherea: 'failed' intertext repetition in Statius' thebaid, or, don't stop me if you've heard this one before / Debra Hershkowitz - Ovid'sTheban narrative in Statius' thebaid / Alison Keith - Statius' hypsipyle: following in the footsteps of Virgil's Aeneid / S. Georgia Nugent - Lacrimabile nomen archemorus: the babe in the woods in Statius' thebaid 4-6 / Joanne Brown - Auferte oculos: modes of spectatorship in Statius' Thebaid 11 / Neil W. Bernstein - Competing endings: re-reading the end of Statius' Thebaid through Lucan / Helen Lovatt - Essential epic: genre and gender from Macer to Statius / Stephen Hinds - Silius' Rome: the rewriting of Vergil's vision / Arthur Pomeroy - Color Ouidianus in Silius' / Richard T. Bruére - Lugendam formae sine uirginitate reliquit: reading Pyrene and the transformation of landscape in Silius' punica 3 /

	Antony Augoustakis - Per uulnera regnum: self-destruction, self sacrifice, and Deuotio in Silius' Punica 4-10 / Raymond Marks - Echoes of the contemporary world in Silius Italicus / Alessandro Mezzanotte.
LC Subjects	Valerius Flaccus, Gaius, active 1st century--Criticism and interpretation.
	Silius Italicus, Tiberius Catius--Criticism and interpretation.
	Statius, P. Papinius (Publius Papinius) - Criticism and interpretation.
	Epic poetry, Latin--History and criticism.
Other Subjects	Silius Italicus, Tiberius Catius.
	Statius, P. Papinius (Publius Papinius)
	Valerius Flaccus, Gaius, active 1st century.
	Epic poetry, Latin.
Form/Genre	Criticism, interpretation, etc.
Notes	Includes bibliographical references (pages 459-499) and index.
Series	Oxford Readings in Classical Studies
	Oxford readings in classical studies.

Fluorescence studies of polymer containing systems

LCCN	2019741871
Type of material	Book
Main title	Fluorescence Studies of Polymer Containing Systems / edited by Karel Procházka.
Edition	1st ed. 2016.
Published/Produced	Cham: Springer International Publishing: Imprint: Springer, 2016.
Description	1 online resource (IX, 302 pages 86 illustrations, 52 illustrations in color.)
	PDF
ISBN	9783319267883
Related names	Procházka, Karel. editor.
Summary	This volume describes the application of fluorescence spectroscopy in polymer research. The first chapters outline the basic principles of the conformational and dynamic behavior of polymers and review the

	problems of polymer self-assembly. Subsequent chapters introduce the theoretical principles of advanced fluorescence methods and typical examples of their application in polymer science. The book closes with several reviews of various fluorescence applications for studying specific aspects of polymer-solution behavior. It is a useful resource for polymer scientists and experts in fluorescence spectroscopy alike, facilitating their communication and cooperation.
Contents	Conformational and dynamic behavior of polymers in solutions - Self-assembly of amphiphilic copolymers in selective solvents - Electrostatically driven assembly of polyelectrolytes - Theoretical principles of fluorescence spectroscopy - Historical perspective of advances in fluorescence research on polymer systems - Fluorescence studies of self- and co-assembling polymer systems - Pyrene-labeled water-soluble macromolecules as fluorescent mimics of associative - Fluorescence correlation spectroscopy studies of polymer systems.
LC Subjects	Spectroscopy.
	Polymers.
	Physical chemistry.
Other Subjects	Spectroscopy/Spectrometry.
	Polymer Sciences.
	Physical Chemistry.
Additional formats	Printed edition: 9783319267869
	Printed edition: 9783319267876
	Printed edition: 9783319800134
Series	Springer Series on Fluorescence, Methods and Applications, 1617-1306; 16
	Springer Series on Fluorescence, Methods and Applications, 1617-1306; 16

Fluorescent methods for molecular motors
LCCN	2019761376
Type of material	Book

Main title	Fluorescent Methods for Molecular Motors / edited by Christopher P. Toseland, Natalia Fili.
Edition	1st ed. 2014.
Published/Produced	Basel: Springer Basel: Imprint: Springer, 2014.
Description	1 online resource (X, 298 pages 88 illustrations, 49 illustrations in color.)
	PDF
ISBN	9783034808569
Related names	Fili, Natalia. editor.
	Toseland, Christopher P. editor.
Summary	This book focuses on the application of fluorescence to study motor proteins (myosins, kinesins, DNA helicases and RNA polymerases). It is intended for a large community of biochemists, biophysicists and cell biologists who study a diverse collection of motor proteins. It can be used by researchers to gain an insight into their first experiments, or by experienced researchers who are looking to expand their research to new areas. Each chapter provides valuable advice for executing the experiments, along with detailed background knowledge in order to develop own experiments.
Contents	Fluorescence and Labelling: How to choose and what to do - Fluorescent biosensors: design and application to motor proteins - Rapid Reaction Kinetic Techniques - Fluorescence to study the ATPase mechanism of Motor Proteins - Use of pyrene labelled actin to probe actin myosin interactions; kinetic and equilibrium studies - Fluorescent methods to study transcription initiation and transition into elongation - Single-molecule and single-particle imaging of molecular motors in vitro and in vivo - Fluorescence methods in the investigation of the DEAD-box helicase mechanism - Use of Fluorescent Techniques to Study the In Vitro Movement of Myosins - Fluorescence Tracking of Motor proteins in vitro - Measuring Transport of Motor Cargos - Measuring two at the same time: Combining

	Magnetic Tweezers with Single-Molecule FRET - Using fluorescence to study actomyosin in yeasts.
LC Subjects	Biological physics.
	Biophysics.
	Cell biology.
	Proteins.
Other Subjects	Protein Science.
	Biological and Medical Physics, Biophysics.
	Cell Biology.
Additional formats	Print version: Fluorescent methods for molecular motors 9783034808552 (DLC) 2014947103
	Printed edition: 9783034808552
	Printed edition: 9783034808576
Series	Experientia Supplementum, 1664-431X; 105
	Experientia Supplementum, 1664-431X; 105

Handbook on flavonoids: dietary sources, properties, and health benefits

LCCN	2011041379
Type of material	Book
Main title	Handbook on flavonoids: dietary sources, properties, and health benefits / editors, Kazuya Yamane and Yuudai Kato.
Published/Created	Hauppauge, N.Y.: Nova Science Publishers, c2012.
Description	xv, 557 p.: ill.; 27 cm.
ISBN	9781619420496 (hardcover)
	ebook
LC classification	QP671.F52 H36 2012
Related names	Yamane, Kazuya.
	Kato, Yuudai.
Contents	Flavonoids: recent insights on their biological action / Salvatore Chirumbolo - Pharmacokinetic variability of dietary phenolic acids and flavonoids in relation to chemical and biological factors / Nabil Semmar, Asma Hammami-Semmar - Modification of flavonoid structure by oxovanadium (IV) complexation: biological effects / Evelina G. Ferrer, Patricia A.M. Williams - Flavonoids and its contribution to a healthier life / Maria do Rosário Bronze, Maria Eduardo Figueira, Elsa Mecha -

Effects of some domestic cooking methods on antioxidant activity, flavonoids, and other phytochemicals content / Irene Dini - Health effects on flavonoids and their relationship in mushrooms / Noboru Motohashi - Dietary flavonoids modulate the oxidative DNA damage induced by N-nitrosamines, heterocyclic amines, and benzo(a)pyrene / Paloma Morales, Ana I. Haza - Impact of conventional and non-conventional technologies applied to obtain fruit products in the flavonoid content and antioxidant capacity of grapefruit / M. Igual ... [et al.] - UV-B radiation: a powerful tool to modulate flavonoid metabolism in tomato fruits / Annamaria Ranieri - Anti-inflammatory properties of dietary flavonoids / A.García-Lafuente, E. Guillamón - Flavonoids: from food and its implication in human health / Montse Rabassa ... [et al.] - Processing of citrus peel for the extraction of flavonoids for biotechnological applications / Munish Puri, Madan Lal Verma, Kiran Mahale - Regulation of intestinal barrier function by dietary flavonoids / Takuya Suzuki - Anti-cancer mechanisms of flavonoids in malignant neuroblastoma / Mrinmay Chakrabarti, Swapan K. Ray - Flavonoid distribution in neglected citrus species grown in the Mediterranean basin / Davide Barreca ... [et al.] - Flavonoids in mushrooms: occurrence, properties, and role of their antioxidant activity / A. Villares - Ginkgo biloba leaves extract (EGb 761) and its specific acylated flavonol constituents increase dopamine and acetylcholine levels in the rat medial prefrontal cortex: possible implications for cognitive enhancing properties of the ginkgo extract / Takashi Yoshitake ... [et al.] - Dietary sources of isoflavones and the methodology used for the analysis / Savithiry S. Natarajan, Devanand L. Luthria .

Other Subjects   Flavonoids.
Notes   Includes bibliographical references and index.

Molecular assemblies: characterization and applications

LCCN	2020034716
Type of material	Book
Main title	Molecular assemblies: characterization and applications / Ramanathan Nagarajan, editor.
Published/Produced	Washington, DC: American Chemical Society, [2020]
Description	1 online resource
ISBN	9780841298873 (ebook other) (hardcover)
LC classification	QD878
Related names	Nagarajan, R. (Ramanathan), editor. American Chemical Society. Division of Colloid and Surface Chemistry, sponsoring body.
Contents	Discovery of monodisperse micelles with discrete aggregation numbers - Supramolecular assembly and mesophase behavior of glycopyranose-derived single-chain amphiphiles - Self-assembly and aggregation studies of simple structural derivatives of stearic acid - Förster resonance energy transfer probing of assembly and disassembly of short interfering RNA/poly(ethylene glycol)-poly-l-lysine polyion complex micelles - Assemblies of hydrophobically modified starch nanoparticles probed by surface tension and pyrene fluorescence - Simple creams, complex structures - Enzyme-triggered nanomaterials and their applications - Characterization of colloidally stabilized latex particles by capillary electrophoresis.
LC Subjects	Supramolecular chemistry. Self-assembly (Chemistry) Colloids. Molecular structure.
Notes	"Sponsored by the ACS Division of Colloid and Surface Chemistry." Includes bibliographical references and index.
Additional formats	Print version: Molecular assemblies Washington, DC: American Chemical Society, [2020] 9780841298880 (DLC) 2020034715

Bibliography

Series	ACS symposium series; 1355

Molecular assemblies: characterization and applications

LCCN	2020034715
Type of material	Book
Main title	Molecular assemblies: characterization and applications / Ramanathan Nagarajan, editor.
Published/Produced	Washington, DC: American Chemical Society, [2020]
Description	xi, 133 pages: illustrations; 26 cm
ISBN	9780841298880 (hardcover)
	(ebook other)
LC classification	QD878 .M64 2020
Related names	Nagarajan, R. (Ramanathan), editor.
	American Chemical Society. Division of Colloid and Surface Chemistry, sponsoring body.
Contents	Discovery of monodisperse micelles with discrete aggregation numbers - Supramolecular assembly and mesophase behavior of glycopyranose-derived single-chain amphiphiles - Self-assembly and aggregation studies of simple structural derivatives of stearic acid - Förster resonance energy transfer probing of assembly and disassembly of short interfering RNA/poly(ethylene glycol)-poly-l-lysine polyion complex micelles - Assemblies of hydrophobically modified starch nanoparticles probed by surface tension and pyrene fluorescence - Simple creams, complex structures - Enzyme-triggered nanomaterials and their applications - Characterization of colloidally stabilized latex particles by capillary electrophoresis.
LC Subjects	Supramolecular chemistry.
	Self-assembly (Chemistry)
	Colloids.
	Molecular structure.
Notes	"Sponsored by the ACS Division of Colloid and Surface Chemistry."
	Includes bibliographical references and index.

Additional formats	Online version: Molecular assemblies Washington, DC: American Chemical Society, [2020] 9780841298873 (DLC) 2020034716
Series	ACS symposium series; 1355

Polycyclic aromatic hydrocarbons: their global atmospheric emissions, transport, and lung cancer risk

LCCN	2019770684
Type of material	Book
Personal name	Shen, Huizhong, author.
Main title	Polycyclic Aromatic Hydrocarbons: Their Global Atmospheric Emissions, Transport, and Lung Cancer Risk / by Huizhong Shen.
Edition	1st ed. 2016.
Published/Produced	Berlin, Heidelberg: Springer Berlin Heidelberg: Imprint: Springer, 2016.
Description	1 online resource (XVI, 177 pages 41 illustrations, 8 illustrations in color.) PDF
ISBN	9783662496800
Summary	This thesis presents a comprehensive analysis of the global health impacts of polycyclic aromatic hydrocarbons (PAHs) in ambient air, conducted on the basis of a high-resolution emission inventory, global chemical transport modeling, and probabilistic risk assessment. One of the main strengths of the thesis is the concentration downscaling process, which provides a linkage between emissions and exposure concentrations at a comparatively high resolution. Moreover, by focusing on individual susceptibility, the thesis proposes an instrumental revision of current risk assessment methodology and argues that, if individual susceptibility were not taken into consideration, the overall risk would be underestimated by 55% and the proportion of highly vulnerable populations would be underestimated by more than 90%.
Contents	Introduction - Research Background - Methodology - Global atmospheric emissions of PAH compounds -

	Global atmospheric transport modeling of benzo[a]pyrene - Global lung cancer risks induced by inhalation exposure to PAHs - Conclusions.
LC Subjects	Air pollution. Atmospheric sciences. Environmental geography. Environmental health.
Other Subjects	Atmospheric Protection/Air Quality Control/Air Pollution. Atmospheric Sciences. Environmental Geography. Environmental Health.
Additional formats	Print version: Polycyclic aromatic hydrocarbons. 9783662496787 (DLC) 2016934444 Printed edition: 9783662496787 Printed edition: 9783662496794 Printed edition: 9783662570258
Series	Springer Theses, Recognizing Outstanding Ph.D. Research, 2190-5053 Springer Theses, Recognizing Outstanding Ph.D. Research, 2190-5053

Pyrene: chemical properties, biochemistry applications, and toxic effects

LCCN	2020688161
Type of material	Book
Main title	Pyrene: chemical properties, biochemistry applications, and toxic effects / Petr Ruzicka and Tomas Kral, editors.
Published/Produced	New York: Nova Publishers, [2013]
Description	1 online resource.
ISBN	9781624172922 (ebook) (hardcover)
LC classification	QP801.P639
Related names	Ruzicka, Petr, 1962- editor. Kral, Tomas, 1968- editor.
LC Subjects	Pyrene (Chemical) Biochemistry.
Notes	Includes bibliographical references and index.

Additional formats	Print version: Pyrene New York: Nova Publishers, [2013] 9781624172915 (hardcover) (DLC) 2012042097
Series	Chemistry research and applications

Pyrene: chemical properties, biochemistry applications, and toxic effects

LCCN	2012042097
Type of material	Book
Main title	Pyrene: chemical properties, biochemistry applications, and toxic effects / Petr Ruzicka and Tomas Kral, editors.
Published/Produced	[Hauppauge] New York: Nova Publishers, [2013]
Description	x, 199 pages: illustrations; 27 cm.
ISBN	9781624172915 (hardcover)
	ebook
LC classification	QP801.P639 P97 2013
Related names	Ruzicka, Petr, 1962-
	Kral, Tomas, 1968-
LC Subjects	Pyrene (Chemical)
	Biochemistry.
Notes	Includes bibliographical references and index.
Series	Chemistry research and applications

Index

A

absorption, vii, 1, 2, 14, 22, 26, 30, 35, 36, 37, 40, 41, 42, 44, 46, 53, 54, 58, 60, 61, 62, 63, 64, 65, 66, 67, 69, 70, 72, 73, 74, 79, 88
analytical, v, vii, 1, 3, 4, 28, 29, 31, 32, 38
anions, 3, 8, 20, 27
application(s), v, vii, viii, 1, 2, 3, 7, 8, 9, 14, 17, 19, 20, 21, 26, 28, 29, 30, 31, 32, 33, 35, 36, 40, 60, 61, 64, 67, 70, 72, 98, 111, 117, 118, 119, 120, 122, 129, 130, 131, 133, 134, 135, 137, 138
aromatic, vii, viii, 1, 2, 3, 5, 6, 9, 10, 14, 23, 30, 35, 37, 40, 41, 58, 71, 77, 80, 81, 93, 98, 104, 122

B

bio imaging, vii, 2, 3, 16, 17, 18, 19, 31, 32

C

cations, 3, 20, 22
characteristic(s), 2, 3, 6, 7, 8, 12, 23, 57, 65, 80, 81, 84, 87, 92, 93, 94, 95
chromophore, v, vii, 1, 2, 11, 35, 36, 37, 66, 82
coal-tar, vii, 35
component(s), 15, 21, 87
conversion, vii, 1, 2, 20, 39, 52, 53, 65, 70
crystal(s), 40, 42, 67, 70, 73, 99, 100, 101, 109, 111, 112, 113, 114, 115

D

discotic, viii, 97, 99, 100, 102, 104, 105, 106, 107, 108, 110, 111, 112, 113, 114, 115
discotic liquid crystals (dlcs), viii, 97, 98, 99, 100, 107, 110, 111, 112, 114, 115
distillation, vii, 35, 36, 37, 98
donor, vii, 1, 2, 6, 17, 53, 54, 55, 57, 58, 59, 61, 62, 65, 71, 81

E

ecotoxicity, 78
emerging pollutants, 78, 87
emission(s), vii, 2, 3, 4, 6, 7, 8, 9, 11, 12, 13, 14, 16, 17, 19, 20, 21, 25, 26, 27, 28, 29, 31, 33, 35, 38, 39, 42, 43, 44, 46, 47, 62, 66, 70, 80, 81, 82, 83, 86, 95, 117, 136
energy, vii, 1, 2, 5, 16, 20, 24, 36, 38, 39, 40, 41, 42, 45, 46, 47, 49, 50, 51, 52, 54, 64, 68, 69, 70, 71, 84, 94, 117, 121, 134, 135
environment(s), viii, 2, 11, 36, 39, 78, 79, 80, 83, 84, 85, 86, 87, 89, 91, 92, 93, 94, 95, 117, 121
explosives, vii, 2, 3, 4, 6, 7, 8, 9, 13, 29, 30

F

fingerprints, 7, 10, 11, 12, 13, 14, 28, 30
fluorescence, vii, 1, 2, 3, 4, 5, 6, 7, 8, 9, 11, 12, 13, 14, 15, 17, 18, 19, 20, 23, 24, 25, 26, 28, 29, 30, 31, 32, 35, 36, 37, 38, 39, 40, 44, 46, 48, 49, 50, 51, 62, 64, 66, 67,

69, 70, 71, 72, 73, 74, 75, 93, 98, 108, 129, 130, 131, 134, 135
forensic(s), v, vii, 1, 3, 4, 7, 8, 9, 20, 28, 29, 30
formations, vii, 1, 2

L

light emitting diodes, 39, 69, 74, 98, 99
liquid crystals, 80, 97, 100, 103, 110, 111, 112, 114
luminescence, vii, 1, 7, 29, 32, 33, 35, 36, 39, 40, 41, 42, 47

M

material(s), v, vii, viii, 1, 2, 3, 7, 10, 11, 12, 13, 14, 20, 28, 29, 30, 32, 35, 36, 40, 41, 42, 44, 46, 47, 49, 51, 53, 54, 60, 65, 67, 69, 70, 71, 73, 74, 77, 78, 81, 82, 83, 84, 85, 86, 87, 88, 92, 94, 98, 101, 102, 105, 106, 107, 108, 110, 111, 112, 117, 118, 119, 121, 122, 125, 127, 129, 130, 132, 134, 135, 136, 137, 138
mesomorphism, 98, 100, 101, 108, 110
micro, v, vii, viii, 7, 17, 28, 46, 77, 78, 79, 80, 84, 85, 86, 87, 88, 89, 90, 91, 92, 93, 95
molecules, vii, viii, 1, 2, 3, 4, 5, 6, 9, 10, 11, 15, 18, 20, 22, 25, 29, 30, 31, 36, 40, 41, 42, 45, 49, 50, 54, 57, 59, 60, 62, 64, 65, 67, 68, 69, 74, 81, 82, 89, 97, 98, 102, 110, 111
monitoring, 4, 78, 92, 93, 117, 118, 120

N

nanoplastic(s), v, vii, viii, 77, 78, 79, 80, 84, 85, 86, 87, 88, 89, 90, 91, 93, 94, 95
nitro aromatics, 3, 4, 5, 8, 9
NLO materials, 35, 60, 68
nonlinear optics, 60

O

optical, vii, viii, 2, 7, 9, 11, 12, 35, 36, 46, 60, 64, 65, 67, 70, 71, 72, 73, 74, 81, 82, 94, 106, 115
optoelectronic, viii, 35, 36, 74, 105
organic, v, vii, viii, 2, 3, 7, 9, 10, 11, 13, 17, 20, 26, 29, 30, 32, 33, 35, 36, 39, 40, 41, 51, 52, 53, 54, 55, 56, 57, 58, 59, 60, 68, 69, 70, 71, 72, 73, 74, 78, 79, 80, 81, 82, 83, 85, 86, 87, 88, 90, 93, 94, 98, 99, 101, 107, 109, 110, 111, 118, 120
organic light emitting devices (OLEDs), 35, 37, 39, 40, 41, 42, 44, 46, 47, 49, 50, 51, 67, 69, 71, 74

P

photophysical, viii, 30, 35, 37, 42, 66, 68, 69, 70, 71, 72, 82, 114
photophysical properties, viii, 30, 35, 42, 66, 82
physicochemical, viii, 77, 78, 81, 82, 85
polarity, vii, 2, 11, 17, 31, 35, 87
polycyclic aromatic hydrocarbons (PAHs), 3, 78, 79, 80, 81, 82, 83, 85, 87, 89, 90, 91, 93, 94, 95, 122, 136, 137
properties, vii, viii, 1, 2, 10, 11, 12, 14, 17, 23, 30, 35, 36, 37, 39, 40, 41, 42, 44, 46, 47, 49, 53, 55, 56, 57, 59, 60, 61, 64, 65, 66, 67, 69, 70, 71, 72, 74, 77, 79, 81, 82, 83, 85, 87, 90, 97, 100, 108, 110, 114, 118, 122, 132, 133, 137, 138

S

self-assembling, v, vii, viii, 97, 110
sensitive, vii, 3, 8, 9, 11, 23, 24, 25, 26, 29, 31, 33, 35, 71, 73
solar cells, 35, 37, 52, 53, 54, 55, 56, 57, 58, 59, 68, 70, 71, 72, 73, 98
structure(s), v, vii, viii, 3, 4, 5, 6, 7, 8, 9, 10, 12, 13, 14, 17, 18, 19, 20, 21, 23, 24, 25, 26, 27, 30, 36, 39, 41, 42, 53, 54, 58, 62, 64, 67, 70, 74, 80, 81, 85, 95, 97, 98,

99, 102, 104, 105, 107, 109, 110, 114, 132, 134, 135
supramolecular, v, vii, viii, 7, 54, 57, 97, 98, 108, 110, 112, 114, 134, 135
synergetic, 88
synthesis, viii, 8, 14, 17, 19, 21, 25, 28, 29, 30, 32, 33, 68, 72, 74, 75, 97, 98, 100, 103, 110, 112, 114, 118

T

toxicity, v, vii, viii, 77, 78, 79, 80, 81, 82, 87, 88, 89, 90, 91, 92, 93, 94
two photon absorption, 35

W

water pollution, 78, 121